SpringerBriefs in Philosophy

More information about this series at http://www.springer.com/series/10082

Robert Francescotti

Physicalism and the Mind

Springer

Preface

This book addresses a tightly knit cluster of questions I've periodically struggled with over the years. One of these questions is: (i) Are mental properties identical with physical properties? An affirmative answer to (i) would seem to secure the truth of *physicalism* regarding the mind, i.e., the belief that all mental phenomena obtain solely in virtue of physical phenomena. If the answer to (i) is negative, then the question arises: (ii) Is there a dependence relation short of identity strong enough to ensure that the mental depends on the physical in the way that physicalism requires? Answering (ii) requires answering two further questions: (iii) What sort of dependence on purely physical phenomena does physicalism require?, and (iv) What is it for a phenomenon to be purely physical? In this book, I try to answer each of (i)–(iv).

How we should answer question (i) depends on how the *multiple realizability* debate should be decided. The idea that mentality is multiply realizable has long been considered a major threat to the view that mental properties are identical with properties of physics, chemistry, or neuroscience. While many philosophers of mind happily abandoned the prospect of psychophysical reduction, others have tried to show that considerations of multiple realizability do not pose any fatal threat. Chapter 1 traces the highlights of the multiple realizability debate over the past few decades, providing an introduction to its many facets, including precise formulations of different varieties of multiple realizability and what their presence does and does not entail. The chapter, however, is not purely expository. In the last third, I present and develop what I consider the best reason for thinking that considerations of multiple realizability do in fact provide irresistible evidence that mental properties are not identical with properties of any of the natural sciences. The question of how physicalists should cope with this purported fact is the topic of the subsequent chapters.

According to the physicalist regarding mentality, mental phenomena obtain solely in virtue of physical phenomena. But how can this view be true if mental properties are not physical? If mental properties are not physical, then it would seem that mentality is not solely a product of the physical. It is here that question (ii) above becomes relevant: Is there a dependence relation short of identity strong

Contents

devastating objection came from the materialist camp itself. Putnam (1967) and Fodor (1974) made popular the idea that mental states are *multiply realizable*. Given the great variety of species that actually exist, Putnam suspected that the very same mental state might be realized in a variety of different neural processes. It seems there is no good reason to think that what instantiates pain in humans is of the same neural type as what instantiates pain in pigs, or ducks, or fish. To suppose that pain is neurally realized in the same way across the wide spectrum of sentient organisms is, as Putnam says, "certainly an ambitious hypothesis" (1967, p. 44).

The idea that mental properties are multiply realizable has been thought by many to show that the identity theory is false. If mentality is multiply realizable at the neural level, then instances of the same mental property can be instances of different neural properties. In terms of the popular *type-token* jargon: tokens of the very same mental state-type might be tokens of different neural state-types. If so, then it would seem to follow that the relation between mental properties and neural properties is one-many rather than one-one. Instances of one mental property may be instances of different neural properties.[5]

Multiple realizability is a threat not just to mind-*brain* identity claims. It poses a threat to identifying mental properties with features of any of the natural sciences. What motivates the thought that the same mental state can be realized by different brain states is the widespread intuition that what makes something a mental state, and the type of mental state it is, is not its inner composition, but the causal role the state plays. Place (1956), Smart (1959), Armstrong (1968), and Lewis (1966) offered causal analyses of mental concepts, which define mental terminology in terms of the typical environmental causes and behavioral effects of the mental state. If what makes something a mental state, and the type of mental state it is, is the causal role it plays, then there are no restrictions on the inner composition of a mental episode. It might be a brain state, a metallic state, or even a hydraulic state (in the Martian that Lewis 1980 describes). The intuition that functional role and not inner structure is what matters to mentality is preserved in the varieties of *functionalism* that evolved from these causal analyses. If this basic functionalist intuition is correct, then it would seem there is a one-many relation, not just between mental properties and brain properties, but between mental properties and any features of any of the natural sciences. Whether brain states, metallic states, or hydraulic states, the one-to-one correlations of the sort needed for true identity claims would seem to be absent.

When focusing on levels of physical structure lower than the level of neurons, the prospect of there being one-one correlations with mentality is even more dim. Neural properties (and their metallic or hydraulic counterparts in computers or Lewisian Martians) are multiply realizable themselves. Processes of the same neural

[5] What is questioned here is the identity theory in its traditional from, the *type-identity* theory of the sort that, e.g., Smart, Place, and more recently, Hill (1991) defended. Multiple realizability is certainly compatible with the *token-identity* thesis. According to the token thesis, each token of a mental state-type is a token of some physical state-type or other, but not necessarily the same physical state-type on each occasion (and perhaps not even a *neural* state-type on some occasions).

type might be constituted in different molecular ways. Molecular properties, in turn, are multiply realizable at the subatomic level, e.g., with carbon dioxide molecules differing from each other in their arrangement of electrons, positrons, and neutrinos. It seems, then, that if mental properties are multiply realizable at one level of physical structure (neural, metallic, hydraulic), then they are also multiply realizable, and to a far greater degree, at lower levels of physical structure, including the molecular, atomic, and subatomic levels. And if so, then mental properties are multiply realizable with respect to properties of the natural sciences generally, from which it would seem to follow that mental properties are not identical with any properties of any of the natural sciences.

It is tempting to draw this anti-reductionist conclusion, and this conclusion has been drawn by many. But is it really a consequence of the multiple realizability of mentality? This question will occupy us for the remainder of the chapter. While many have argued that this failure of identity is not a consequence of the way in which mentality is multiply realized, we shall see by the end of this chapter that despite efforts to prove otherwise, mental properties really are distinct from those of physics, chemistry, and biology.

1.2 Multiple Realizability and Reductionism

In his famous support of the autonomy of psychology, Fodor (1974) appeals to Nagel's (1961) influential account of theoretical reduction. On the Nagelian account of reduction, a theory T is reducible to a theory T* only if each statement of T is *derivable* from a statement of T*. Typically, the vocabulary of the reduced theory T is not synonymous with the vocabulary of the reducing theory T*. So generally the laws of T will be derivable from the laws of T* only with the help of *bridge principles* linking the predicates of T to those of T*. Fodor supposes that these bridge principles will be true *biconditionals*—of the form, '$F_1x \leftrightarrow G_1x$' and '$F_2y \leftrightarrow G_2y$', where the F-predicates are the predicates of one of the theories and the G-predicates are those of the other theory. Fodor claims, moreover, that successful reduction of T to T* requires that these biconditional bridge principles are true as a matter of *nomic necessity*. For "bridge laws are *laws*" and "a law must succeed with the counterfactuals" (1974, p. 104), where counterfactuals are conditionals whose truth or falsity depends on what obtains in other possible worlds with the same laws of nature. Now, where the reduction of psychology to the natural sciences is concerned, Fodor finds it doubtful that there are true biconditional bridge principles, and quite improbable that these biconditionals are true as a matter of nomic necessity. As he remarks, "it seems increasingly likely that there are nomologically possible systems other than organisms (namely, automata) which satisfy natural kind predicates in psychology, and which satisfy no neurological predicates at all" (p. 105). He concludes that psychology is not reducible to neurophysiology or any other natural science.

With his talk of 'psychological' theories, Fodor presumably means not just theories that actually have been endorsed by psychologists or those that will end up being popular in their field, but theories of any *mentalistic* variety, including "folk psychology." To avoid focusing only on the science of psychology, let's use the term 'mental' instead of 'psychological' and represent Fodor's line of argument as follows:

(1) Mental properties are multiply realizable with respect to properties of the natural sciences.
(2) If mental properties are multiply realizable with respect to those of the natural sciences, then there are no nomically necessary biconditionals linking the former to the latter.
(3) If there are no nomically necessary biconditionals linking mental properties to those of the natural sciences, then mentalistic theories are not reducible to those of the natural sciences.

Therefore, (4) mentalistic theories are not reducible to those of the natural sciences.

Let's call this the "Multiple Realizability Argument for Non-Reductionism," or "MR_{NR}" for short.

While MR_{NR} has been one of the main reasons for the immense popularity of *non-reductionism* in the philosophy of mind for the past few decades, it has not been without its share of criticism.

1.3 Reduction without Nomic Equivalence

A common reply to MR_{NR} is to object to premise 3. It has been shown that there are widely recognized cases of reduction in which the reduced property happens to be multiply realized, from which it is concluded that multiple realizability is no threat to reductionism. For example, Bickle mentions that while temperature is considered a physical phenomenon,

> [w]hen you think in terms of the velocity and momentum of each individual molecule, you will see that there are an *indefinite* number of ways for a given aggregate of molecules to realize any given temperature (1992, p. 53),

and Robert Richardson reports that

> oxygen has three isotopes—0–16, 0–17, and 0–18—varying not with respect to charge, but in the number of neutrons—eight, nine, and ten, respectively—in the nucleus.... The existence of isotopic variants is a common feature in chemical "elements".... Even in the reduction of chemistry to physics, what is one type at a higher level (chemistry) is segregated into several importantly different types at immediately adjacent levels (1982, p. 132).

If multiple realizability is compatible with reductionism, as these examples seem to show, then where and how exactly does MR_{NR} fail? Fodor assumes that reduction requires the presence of true *biconditionals* linking the properties of the reduced theory to those of the reducing theory. Yet, as Richardson indicates, reduction "requires no more than a mapping *from* lower *to* higher level types and *not* a mapping from higher to lower level types" (1979, p. 548). The reduction of theory T to T* requires the *derivability* of the statements of T from those of T*, which demands only true *one*-way conditionals linking the predicates of T to those of T*; true *bi*-conditionals are not required. In fact, as Richardson notes, Nagel himself claims that "the linkage between *A* [a term in the secondary science] and *B* [a term in the primary science] is not necessarily biconditional in form, and may for example be only a one-way conditional: If *B*, then *A*" (Nagel 1961, p. 355).[6]

Moreover, as is widely recognized, it is not even necessary that the reduced theory is itself derivable from the reducing theory. As Paul Churchland reminds us, "relativistic mass is not identical with Newtonian mass, nor even coextensive with it, even at low velocities," and yet "the reduction of Newtonian by Einsteinian mechanics is a paradigm of a successful reduction" (1985, p. 10). Churchland claims that T being reducible to T* requires no more than that T* (along with limiting assumptions and boundary conditions) entails a set of principles, *I*, that is an adequate isomorph of T. On this model, there is no need for conditionals, bi- or even one-way, linking the terms of T to those of T*, for on this model, it is an *image* of T rather than T itself that is derivable from T*.[7] As Churchland notes, this approach to intertheoretic reduction is developed by Hooker (1981). And inspired by Hooker's insights, Bickle (1998) provides a detailed elaboration of the isomorph model of reduction and calls it "new wave" reductionism.

Whether we endorse Bickle's elaboration of Hooker's idea or some other version of the isomorph model, we allow reduction even without the derivability of T from T*. And even if we think of reduction as requiring the derivability of T itself, there is Richardson's point that this derivability requires only one-way conditionals. In either case, the consequence for psychophysical reduction is the same. Even if mental properties are multiply realizable, and even if this entails that there are no true biconditionals connecting the statements of mentalistic theories with those of the natural sciences (as premise 2 of MR_{NR} maintains), mentalistic theories might still be reducible to theories of the natural sciences after all. So if Churchland, Bickle, and Richardson are correct about what reduction does and does not require, then premise 3 of Fodor's anti-reductionist argument as formulated in the previous section is false; theory reduction does not require biconditional bridge principles.

In response to this objection to premise 3, one might insist that without nomically necessary biconditionals linking the properties of T to those of T*, T simply does not warrant being called "reducible." This line of response is not very promising since the word 'reduction' is far from univocal. Not only are there many different ways, and arguably perfectly legitimate ways, to understand theoretical reduction in the sciences, there are other varieties of reduction that philosophers

---

[6] This is quoted by Richardson (1979, pp. 548–549).

[7] See, also, Churchland (1979, Sect. 11).

often mention.[8] Given the variety of ways the word 'reduction' has been used in the philosophical literature, it would seem unfair to accuse views that do not require nomically necessary biconditionals as relying on an illegitimate understanding of reduction.

Yet, while there is good reason to reject MR_{NR}, there is a closely related argument that is not so easily refuted, one that focuses on the special brand of reduction that requires property-*identity*.

1.4 The Multiple Realizability Argument for Non-Identity

The goal of a philosophical theory of mind is to reveal the nature of mentality, and this obviously involves, among other things, revealing what exactly the various mental properties *are*, i.e., specifying that with which the various mental properties are *identical*. Now, on any plausible account of property identity, if property F is identical with property G, then it is nomically necessary that for any individual x, Fx ↔ Gx. Thus, while some varieties of reduction do not require nomically necessary biconditional bridge principles, there is one highly significant brand of reduction that does—the brand of reduction that aims to specify property-identities.[9] It is arguable that property-identity, and identity in general, requires a brand of necessity stronger than nomic necessity; it is arguable that F = G only if it is *metaphysically* necessary that for any x, Fx ↔ Gx. However, revising MR_{NR} to escape the reductionist reply described in 1.3 does not require that we appeal to metaphysical necessity. It requires no more than replacing the reference to reduction with talk of identity.

(1) Mental properties are multiply realizable with respect to properties of the natural sciences.
(2) If mental properties are multiply realizable with respect to those of the natural sciences, then there are no nomically necessary biconditionals linking the former to the latter.
(3) If there are no nomically necessary biconditionals linking mental properties to those of the natural sciences, then the former are *not identical with* the latter.

Therefore, (4) mental properties are *not identical with* those of the natural sciences.

[8] See, for example, Searle's (1992, pp. 112–116) description of some of the many different varieties of reduction.

[9] This brand of reduction warrants the title "ontological reduction." As Clapp indicates, "[i]n the sense in which *reduction* is relevant to the mind-body problem ..., the reduction of one theory T_1 to another theory T_2 collapses the ontological commitments of T_1 to those of T_2" (2001, p. 114). Also, as Enc informed us, "one of the objectives of scientific reduction is a reduction in the ontology of things and of properties. And this will not be achieved unless type-type identities are established between the two sciences. And clearly these identities cannot be established unless bridge laws express coextensionalities" (1983, p. 280).

Let's call this argument the "Multiple Realizability Argument for Non-Identity," or "MR_{NI}" for short.

The reductionist response to MR_{NR} does not threaten MR_{NI}, for the fact that there are many legitimate senses of 'reduction', some of which do not require true biconditionals, is no objection to the new premise 3.[10] It seems that the only option for refuting the new argument, MR_{NI}, is to reject either 1 or 2. To help decide which of these premises should be rejected, if either, it will be useful to distinguish between some different varieties of multiple realizability.

1.5 Mild and Robust Multiple Realizability

Let us say that a property F is *mildly* multiply realizable with respect to a class of properties *C* if and only if

> there are members, G and H, of *C* such that G ≠ H and it is nomically possible that: an instance x of G instantiates F and an instance y of H instantiates F.[11]

For any particular sample of gold, there is another sample that differs somewhat in physical detail, perhaps containing a few more Au atoms, or by differing in the number and kind of other constituent atoms. So what instantiates the property of being gold in one case might differ chemically from what instantiates that property in another case. So gold is mildly multiply realizable with respect to the class of chemical properties.

But this fact, obviously, does not prevent gold from being nomically coextensive (and even identical) with a chemical kind. Since Au is itself a chemical kind, there is a chemical feature common to *all* possible instances of gold (at least all nomically possible instances), and since that chemical feature (being comprised of Au) is also present *only* in those instances, there are nomically necessary biconditionals linking the property of being gold with being comprised of Au. So being multiply realizable in the mild sense with respect to some class of properties does not entail being nomically inequivalent to each of those properties. As Polger (2004) mentions,

[10] Even if one thought, with Bickle, that reductionism "is a claim about how distinct scientific endeavors relate to one another… not explicitly an account of how phenomena relate—objects, entities, properties, processes, states of affairs, and the like—nor an account of the relations that obtain among them" (2010, p. 256), one who cared about the nature of mentality would still have to contend with MR_{NI}.

[11] Funkhouser defends a definition equivalent to this. He proposes that a property-type X is multiply realizable relative to science Y iff there are possible instances of X that "exactly resemble one another with regard to their *X*-ness," but "their *Y*-realization bases do not *exactly* resemble each other with regard to their *Y*-ness" (2007, p. 476). On this account, mental properties are multiply realizable at the neural level precisely because instances that exactly resemble each other mentally (i.e., are instances of the very same mental property) do not exactly resemble each other neurally (i.e., are instances of different neural properties). What I am calling "mild" multiple realizability is also similar to what Polger (2004, p. 6) labels "weak" multiple realizability.

carburetors are multiply realizable in that "[t]here are many manufacturers of carburetors, which are made of various materials and of various designs," yet "there is a single structural feature that all carburetors have: they all have venturi" (pp. 19–20). So while carburetors are multiply realizable with respect to physical properties in the mild sense defined above, there is still a physical feature that is both necessary and sufficient for being a carburetor—i.e., being a device for mixing air and fuel according to the venturi principle.

Recall the cases of scientific reduction mentioned in Sect. 1.3. Bickle is right to point out that "there are an *indefinite* number of ways for a given aggregate of molecules to realize any given temperature" (1992, p. 53). Yet, as Endicott reminds us, "temperature is not multiply realized with respect to *all* the physical properties because temperature is itself a property of physical science (thermodynamics), and no property is multiply realized with respect to itself" (1989, p. 218). Since temperature itself is a physical property, there is a physical property that is exemplified if and only if the property of having a certain temperature is exemplified. Also recall Richardson's isotope example. It is true that the three isotopes of oxygen differ in the number of neutrons. But despite this difference there is a commonality. There is a common subatomic feature that makes the three isotopes of oxygen isotopes of the same atom, the property of *having the same number of protons*. In both cases, there is mild multiple realizability but not without true biconditionals linking the property to a property of physical science.

How does this relate to minds and brains? Kim wrote, "the fact that two brains are physico-chemically different does not entail that the two brains cannot be in the 'same physico-chemical state'" (1972, pp. 189–190). Of course, two brains can differ while having some neural similarities. But there is a further point that Kim wishes to make, i.e., two individual processes might differ neurologically while still having some neural similarities. Adams makes the same point: "just because two systems are different kinds of stuff does not mean that they do not share some identical property-kinds" (1979, p. 158). It is clear, then, that a mental property's being multiply realizable in the mild sense with respect to neural properties allows that there is a neural property that accompanies the mental property with nomic necessity. So it seems that the standard examples from the philosophical literature (e.g., pain instantiated in one type of neural state in humans, another in dogs, another in octopi) shows that mental properties are multiply realizable in the mild sense, but they do not show that they are multiply realizable in a way that entails premise 2 of MR_{NI}.[12] Even imagined cases in which non-biological entities bear mentality (e.g., human-made metallic machines or Lewis' Martian) fail to establish premise 2. Even if an individual with mentality is wholly inorganic, for each of its mental properties, there might be a property describable by one of the natural sciences (physics or chemistry presumably) that it shares with us, and one that qualifies as nomically necessary and sufficient for the presence of the mental property.

---

[12] I stress this point in my (1997).

It is clear, then, that some stronger brand of multiple realizability is needed to establish the absence of nomically necessary biconditionals. Let's try the following. Suppose that property F is multiply realizable with respect to a class of properties *C* if and only if

> there are members, G and H, of *C* such that G ≠ H and it is nomically possible that: an instance x of G instantiates F and an instance y of H instantiates F, *and there is no property I of C that x and y both instantiate.*

This formulation requires that there are different instances of F that have nothing in common *C*-wise—e.g., different instances of a mental property that have nothing in common physically, chemically, or biologically. So if mental properties are multiply realizable in this strong sense with respect to those of the natural sciences, then thanks to the italicized clause there are no true biconditionals linking the former to the latter, just as premise 2 of MR_{NI} claims.

Nonetheless, the formulation above makes multiple realizability far too difficult to come by. Suppose that substance dualism is false, and suppose it is false as a matter of nomic necessity; suppose that the laws of nature that actually obtain prevent mentality from being realized in some wholly non-physical way. In that case, every instance of any mental property instantiates the highly general physical properties of *occupying physical space* and *inhabiting the physical universe* as a matter of nomic necessity, which thanks to the italicized clause entails that no mental property is multiply realizable with respect to physical properties in the sense above.

Here is a type of more than mild multiple realizability that is more likely to obtain. Let's say that a property F is *robustly* multiply realizable with respect to a class of properties *C* if and only if

> there are members, G and H, of *C* such that G ≠ H and it is nomically possible that: an instance x of G instantiates F and an instance y of H instantiates F, and there is no property I of *C* that x and y both instantiate *and that is common only to instances of F.*[13]

For any mental property M, *occupying physical space* and *inhabiting the physical universe* are not peculiar to M. So this new formulation allows the multiple realizability of mental properties despite some general physical properties that each of their instances share. At the same time, if mental properties are multiply realizable with respect to properties of the natural sciences in the robust sense above, then it is assured that there are no nomically necessarily biconditionals linking them to any properties of the natural sciences.[14]

---

[13] This formulation is equivalent to Endicott's (2005) definition of multiple realizability.

[14] Aizawa and Gillett (2009, 2011) offer the following analysis: A property G is multiply realized *if and only if* (i) under condition S, an individual *s* has an instance of property G in virtue of the powers contributed by instances of properties/relations F_1-F_n to *s*, or to *s*'s constituents, but not vice versa; (ii) under condition S* (which may or may not be identical to S), an individual *s** (which may or may not be identical to *s*) has an instance of property G in virtue of the powers

One might wish to add to our formulation Shapiro's (2000) requirement that a property or kind is multiply realizable only if there are different ways to causally implement the function that defines that property/kind. Shapiro proposes that "two realizations of a kind *T* are in fact different kinds of realizations of *T* only when they differ in their causally relevant properties, that is, the properties by which they contribute to the capacity, purpose, goal, and the like that serves to individuate *T* as the kind that it is" (2000, p. 646). To use his example, it might be thought that the steel corkscrew and the aluminum one are different realizations of the kind *corkscrew*. But since the properties causally relevant to removing corks are the same in both cases, Shapiro denies that this is a genuine case of multiple realizability. On the other hand, the double-lever corkscrew and the waiter's corkscrew achieve the goal of removing corks in different ways; so they count as different realizations of the *corkscrew* kind.

It is arguable, however, that a difference in realization does not require a difference in the method of realization. Suppose that some functionalist account of mentality is correct, and that two physical properties realize the same mental property by virtue of playing the causal role definitive of the mental property. Since the mental property is realized in both cases in virtue of the same causal role being played, the causally relevant properties are the same in this case. So on Shapiro's account, this would not qualify as a case of genuine multiple realizability. But it certainly seems that it does qualify, especially if the difference in realizers is as great as the difference between Martian pain and human pain. Indeed, the very notion of multiple realizability was introduced precisely to allow for the possibility of such cases: the same functional role being played by diverse physical realizers. And even if one wishes to reserve the label 'multiple realization' for cases where there is a causal difference, the type of variable realization that poses a threat to psychophysical identity claims is not restricted to cases of causal difference. Even if different *C*-properties, G and H, give rise to F in exactly the same way, without a *C*-property common to all and only instances of F, F is not identical with any *C*-property. So in considering the type of variable realization that threatens psychophysical identity claims, we should not confine our attention to cases in which the realized property is instantiated in causally distinct ways.[15]

(Footnote 14 continued)
contributed by instances of properties/relations F^*_1-F^*_m of *s** or *s**'s constituents, but not vice versa; (iii) F_1-$F_n \neq F^*_1$-F^*_m, and (iv), under conditions S and S*, F_1-F_n of *s* and F^*_1-F^*_m of *s** are at the same scientific level of properties. Despite its merits, their detailed account does not preclude an F-level property which F_1-F_n and F^*_1-F^*_m share, and which is nomically necessary and sufficient for the realized property G.

[15] Shapiro contends that those who support the multiple realizability of mentality face a dilemma. Either the realizing kinds differ in their causally relevant properties or they do not. If they do not, then we do not have a case of genuine multiple realizability. If they do, then there is reason to suspect that they are not realizing the same mental property. For if the realizers "have no or only few causally relevant properties in common, then there are no or just a few laws that are true of both of them" (2000, p. 649), in which case, it is arguable that we have here the realization of genuinely different functional kinds. However, if I am right, one can and should resist at least the first horn of this dilemma.

One might wonder how Gillett's (2003) distinction between *flat* and *dimensioned* views of realization relates to our formulation of robust multiple realizability. On the flat account the realized property and the realizer are instantiated in the same individual; e.g., the realized property of *being a corkscrew* is a property of the very same object that has the realizer property, *being an aluminum object of such-and-such physical dimensions*. The dimensioned account, on the other hand, allows realized and realizer properties to be instantiated in different individuals. It allows, for instance, that the realizer property is instantiated not in the object that has the realized property, but in the *parts* of that object. Gillett borrows Shapiro's corkscrew example to illustrate the difference between the two accounts. On the flat account, "the properties/relations of the constituent steel molecules and aluminum atoms cannot be the basis for a multiple realization," since these properties/relations "are not instantiated in the corkscrew, but in its constituents" (p. 597). However, since the dimensioned account allows properties of constituents to be realizers, we can plausibly view the case of the steel and aluminum corkscrew as an instance of multiple realization even if we accept the causal difference requirement. "For with one corkscrew the properties/relations of steel molecules result in the powers individuative of the property of removing corks, while in the other corkscrew the distinct properties/relations of aluminum atoms play this role" (p. 598).[16]

I will not discuss here which account is to be preferred, the flat or the dimensioned view,[17] but only note that both are perfectly compatible with multiple realizability in the robust sense described above. Recall:

> there are members, G and H, of *C* such that G ≠ H and it is nomically possible that: an instance x of G instantiates F and an instance y of H instantiates F, and there is no property I of *C* that x and y both instantiate and that is common only to instances of F.

This formulation allows that G and H are properties of the same individual that has property F, but this same-subject realization is not required. Suppose that F is the property of desiring pizza and that F is instantiated at the atomic level in manner P on one occasion and in manner P* at the atomic level on some other occasion. Also suppose that G is the property of atomic constituents distributed and arranged in manner P and that H is the property of atomic constituents fitting profile P*. In this case, an instance of G is an instance of F and an instance of H is an instance of F, even though G and H are properties of the constituents (the constituent atoms) while F is a property of the whole (the pizza craver herself). And assuming there is no

---

[16] Gillett argues that Shapiro's reasoning against the multiple realizability of mental properties is ineffective since it presupposes the flat account of realization whereas the non-reductionist's appeal to multiple realizability is based on the dimensioned view.

[17] Although see Polger's (2007, 2008) critique of Gillett's dimensioned account. Also see Polger and Shapiro (2008).

property at the atomic level common to all and only nomically possible instances of F, F is multiply realizable in the strong sense at the atomic level on the dimensioned view.

It's not clear whether we should accept premise 2 of MR_{NI} and that's because its truth depends on the brand of multiple realizability at issue. We saw that with the mild variety, 2 is false, but it's true with the robust brand. So let's focus on the robust brand and refine the first two premises of MR_{NI} as follows:

(1) Mental properties are *robustly* multiply realizable with respect to properties of the natural sciences

and

(2) If mental properties are *robustly* multiply realizable with respect to those of the natural sciences, then there are no nomically necessary biconditionals linking the former to the latter.

It seems fairly clear that mental properties are multiply realizable in the mild sense with respect to properties of physics, chemistry, and biology. However, it is not at all clear that they are multiply realizable in the robust sense, for as we have seen, any lower-level differences may be accompanied by lower-level similarities. So now the first premise of MR_{NI} becomes the major source of controversy. Objections to our new premise 1 will be discussed in Sect. 1.7. But first let's consider a popular response to multiple realizability arguments which might be used to question our new premise 2 or the original premise 3 of MR_{NI}.

1.6 Species-Specific Property Identities

That mental properties can be realized differently in different populations does not prove that there are no true biconditionals linking mental properties to properties of physics, chemistry, or biology. For there might be true psychophysical biconditionals that are *relativized to populations*. Even if M is multiply realizable in the robust sense across species, it might be that for any species S, there is a property P of the natural sciences, such that *M* ↔ *P for species S*. And these species-specific biconditionals might license species-specific identity claims: *M* = *P for species S*. In the literature on multiple realizability, the appeal to species-specific reductions has a long history. Lewis (1969, 1980) and Kim (1972) have endorsed species-specific identities. Churchland (1979, 1982, 1984) did as well, noting that the appeal to domain-specific reductions is standard scientific practice. Churchland points out that temperature has been "cited as the paradigm of a successfully *reduced* property" even though temperature is mean kinetic molecular energy only for a gas, "where simple particles are free to move in ballistic fashion." In a solid, "temperature is realized differently, since the interconnected molecules are confined to a variety of vibrational motions," in a plasma, "temperature is something else again," and "even a *vacuum* has a so-called 'blackbody' temperature" (1984, p. 41). So contrary to the new premise 2 of MR_{NI}, even if mental properties are robustly multiply realizable with respect to those of the natural sciences, there might still be nomically necessary biconditionals of the species-specific variety, linking the former to the latter. And against premise 3 of MR_{NI} one can appeal to identity claims of the form $M = P$ *for species S* despite the falsity of the unrelativized biconditionals.

The appeal to species-specific identity claims is not without objections. There is the worry that with species-specific identity claims, we are no longer capturing the nature of the mental property itself. Pain is supposed to be a mental property that we share with dogs, birds, and octopi, but if we are told only what pain-in-birds is or pain-in-dogs, then we are not being told what that general feature is in virtue of which we are all in some sort of pain. The appeal to species-specific identities, as Block puts it, "sidesteps the main metaphysical question: 'What is common to the pains of dogs and people (and all other pains) in virtue of which they are pains?'" (1980, pp. 178–179).

A related worry is that if the best we can find are species-specific identity claims, then it seems that the common mental property, M, with which we started simply does not exist. Suppose M is to be identified with physical property P_1 in species S and physical property P_2 in species S*. Since P_1 and P_2 are different properties, it follows that the mental property instantiated in the members of S is a different mental property than the one instantiated in the members of S*. In that case, the members of S and S* would not really be sharing the same mental property. Churchland's temperature example only reinforces the worry that with species-specific reduction, the property originally thought to be shared is not a common feature after all; the phenomenon called "temperature" in a gas is not the same phenomenon as that labeled "temperature" in a liquid, which is not the same as that called "temperature" in a solid. So with only species-specific psychophysical identity claims available, it seems that there is no genuine sharing of mental properties across the different species. The properties that we thought were shared would not really exist.[18] This seems to be an implausible consequence. Granted, the qualitative character of the pain felt by some bird differs from the qualitative character of your most recent pain episode, and the content of a dog's belief that it's raining is different from the content of my belief that it's raining, but it certainly seems that there are some general mental properties (e.g., feeling pain of some sort, and having some belief about the weather) that we share.

There is another, more obvious problem with the appeal to species-specific psychophysical identities. Even within the same species, there is no guarantee that each mental property is realized in the exact same way. For all we know, it might be that what neurally underlies my belief that $2 + 3 = 5$ is not the same as what realizes that belief in you. Lewis acknowledges this possibility—that pain "might even be one brain state in the case of Putnam, another in the case of Lewis" (1969, p. 25).

---

[18] Perhaps this is what Endicott has in mind when he says that "we should not look on the NRS [the 'narrow reductive strategy'] as a reductive strategy at all, but as a form of *eliminativism*" (1993, p. 310).

Kim also admits that to generate psychophysical laws, "biological species may turn out to be too wide; individual differences in the localization of psychological functions in the brain are well known" (1989, p. 38). Even more worrisome for the appeal to species-specific identities is that multiple realizability might occur within the life of the very same individual.[19] Given the plasticity of the brain, it would seem that intra-personal multiple realizability actually obtains quite often. There is much empirical evidence that with brain damage, head trauma, changing task demands, and ordinary development, the neural regions and machinery that serve a certain cognitive function at one time in an individual's life might differ from what plays that cognitive role at another time in that individual's life.[20]

In response to the objection regarding intra-species and intra-personal multiple realizability, one might propose identity claims that are relativized to individuals, and to individuals at times. But then, as Endicott points out, it seems we are no longer talking about *type*-identities: "the whole rationale for talking about types or properties is that they are *general* features, instantiated or instantiatable on different occasions," which is not true of something like *M for individual x at time t*; "the more narrow we make our properties, the less general they become, and the less general they become, the more they become like particulars" (1993, p. 317). So with extreme relativization, type-identities are turned into token-identities.

It seems that the most promising way to refute MR_{NI} is not to question 2 or 3, but to question the revised premise 1—i.e., the claim that mental properties are multiply realizable in the robust sense with respect to properties of the natural sciences.

1.7 Empirical Evidence against Robust Multiple Realizability

An article by Bechtel and Mundale (1999) has deservedly attracted much attention. A better understanding of actual neuroscientific research, they argue, reveals that multiple realizability is not as rampant as many philosophers have supposed. One major point of their discussion is that many who accept the multiple realizability argument against psychophysical reduction (including psychophysical identity claims) do so because they individuate mental properties more coarsely than they individuate neural properties. By using a coarse grain in classifying mental states and a fine grain in distinguishing between brain states, one is bound to find many mental similarities amidst the neurological differences. It is inevitable that we will find a coarsely individuated mental property corresponding to several finely

[19] Horgan (1993, p. 308) makes this point regarding intentional states.

[20] For details about neural plasticity see, for example, Endicott (1993), Buonomano and Merzenich (1998), Stein and Hoffman (2003), Draganski et al. (2006), Wieloch and Nikolich (2006), and Richardson (2009). Although, Polger (2009) argues that neural plasticity does not provide definitive evidence of multiple realizability.

individuated neural properties. However, if one individuates mental properties with the same grain used to individuate neural properties, then it is more likely that one-one correspondences will be found.

They are, of course, exactly right about this. There are ways of individuating mental kinds and neural kinds that yield a one-many correspondence between the two classes, which shows that mental kinds are *mildly* multiply realizable with respect to neural kinds. But it is compatible with this that there is a way to individuate mental and neural kinds that yields a one-one correspondence. At the same time, the fact that one can individuate properties more or less coarsely does not by itself show that there actually is a method of individuating that reveals a one-one mapping of mental kinds onto neural kinds. Is there any good reason to believe there actually is a method of carving that yields such a correspondence?

Bechtel and Mundale argue that there is evidence to support this belief coming from actual neuroscientific practice. They point out that it is "precisely on the basis of working assumptions about commonalities in brains across individuals and species that neurobiologists and cognitive neuroscientists have discovered clues to the information processing being performed" (1999, p. 177). They report that Korbinian Brodmann's famous brain maps were developed comparatively, by considering brain regions in many different species, including humans, marmosets, lemurs, hedgehogs, and ground squirrels, and from these brain maps Brodmann concludes (as Bechtel and Mundale report) that there is "similarity in the overall patterns of parcellation, constancy in broader regions (pre-central, granular frontal region, etc.) across species, and persistence of individual areas" (pp. 180–181). Bechtel and Mundale also mention Franz Gall's attempt to map psychological functions onto the brain on the basis of cranial protrusions and indentations, pointing out that Gall "demarcated cranial areas in light of psychological functions, insuring that function was built into the characterization of the region" (p. 184). In addition, there is David Ferrier's brain stimulation research in the late nineteenth century, which was highly comparative, with experiments on a variety of mammals (including macaque monkeys, dogs, jackals, cats, rabbits, and guinea-pigs) and a few non-mammals. Bechtel and Mundale also describe the comparative nature of recently developed techniques such as PET and fMRI for studying brain activity (pp. 189–190). Since there are differences between individual brains, researchers using these techniques have found it necessary to refer to a common atlas to designate which areas are active, and this requires a mapping of different brains onto the common atlas.

In the studies Bechtel and Mundale report, (a) neuroscientific research is guided by the assumption of brain commonalities within and between species. There is also the crucial point that in these studies (b) brain areas and processes are individuated in terms of the cognitive role they play. And with an emphasis on cognitive contribution in individuating brain regions, it is not implausible to think that (c) the grain of individuating brain processes will in many cases approximate the grain with which the relevant mental processes are individuated.

It seems that Bechtel and Mundale have clearly achieved what they identify as their "primary concern"—i.e., to disprove "the implication drawn from the multiple realizability argument that information about the brain is of little or no relevance to understanding psychological processes" (p. 176). No one should believe that information about the brain is of little relevance to understanding psychology; it is clear that brain research of the sort Bechtel and Mundale mention has greatly enhanced the understanding of cognitive functions.[21] It is also arguable that their detailed analysis of neuroscientific research shows that (d) multiple realizability (of the robust variety) is not quite as pervasive as philosophers have thought.

While granting (a–d), the question remains whether Bechtel and Mundale have shown that there's room for optimism about a one-one mapping of mental states onto brain states. It is doubtful that they have shown this. Firstly, to justify accepting mind-brain identity claims (and not just the species-specific identity claims discussed in 1.6), one needs to show that with respect to all instances of the same mental property *across all species*, there is a way to individuate brain states that yields a one-one correspondence with mental states. However, as Sungsu Kim (2002) and Richardson (2008) point out, Bechtel and Mundale place too much emphasis on *homologues*, species sharing a common lineage. Richardson notes that "[w]hat we would expect, from an evolutionary standpoint, is for multiple realization to show up when we look to more distant forms. Homology will obscure the range of possibilities" (2008, p. 534). Shapiro (2008) makes the additional point that by focusing on homologous cases, whatever differences in brain structure are found may be explained in terms of slight differences in the property being realized. For example, "hummingbirds and eagles share flight properties in virtue of possessing wings that are similar, and whatever differences there are in their wings explain only differences in their flight properties" (p. 524). However, if we consider non-homologous cases, e.g., hummingbirds and bats, we are more likely to find clearer instances in which differences in structure underlie the very same property. Thus, as Kim informed, it is "homoplasies, not homologies, that can provide evidence for or evidence against multiple realizability" (2002, p. 610).

So even with points (a–d) above, which Bechtel and Mundale have established, it remains doubtful that there are the interspecies mind-brain correspondences needed to support identity claims. In fact, Bechtel and Mundale admit that the very researchers they mention found evidence that their comparative approach does *not* apply to all species. "While Brodmann also drew attention to differences between species, his primary concern was to show commonalities in brain regions, not differences" (1999, p. 181), and Ferrier's stimulation "studies on pigeons, frogs, and fishes" showed "very few locations where stimulations could elicit any responses," unlike his experiments on mammals (p. 185).[22]

---

[21] In Sect. 4 of their essay, they describe in detail how studying the structure of the brain has greatly contributed to our understanding of visual processing.

[22] For additional objections to the reasoning of Bechtel and Mundale (1999) see Aizawa's (2009). Aizawa argues against two premises in what he calls the "Central Argument" of Bechtel and Mundale. Aizawa objects to the idea that brain taxonomy requires an appeal to psychological function, and he argues against the premise that if psychological functions were multiply realized, then brain taxonomy would have to be carried out independently of psychological function and

On behalf of Bechtel and Mundale, one might argue that whenever there are interspecies neural differences for what seems to be a common mental feature, what is exemplified is not really the same mental feature after all. Polger contends that "*species-specific* variation of mental states is to be expected," a variation of the mental states themselves, not just the underlying neural or otherwise physical states. "We expect only very similar creatures to have very similar sensations. To the extent that a creature differs from us, so will its experiences" (2004, p. 15). So, Polger contends, given the physical differences, there is reason to suspect that pain "in human beings may not be exactly similar to pain in dogs, dolphins, or Martians" (p. 15). And Couch (2004) points out that in many cases of apparent multiple realizability, it is not implausible to think that the realized property is actually not the same property; e.g., one can appeal to functional differences to argue that the visual properties realized by primate eyes are different from those realized by the octopus eye.[23,24] Granted, it is not unreasonable to expect genuine mental differences (and not just neural differences) in the way that some feature M (e.g., vision, hunger, or pain) is exemplified in two different species. However, it seems that in many such cases, there is some broadly individuated mental property—M itself (hunger or pain in general)—that is had in common despite the neural differences.[25]

The type-identity theorist might still hold fast, insisting that in every case where it seems that there is no neural condition common to the various instances of the same mental property, there really is some brain feature common to all of them, and a brain feature that is also sufficient for the presence of the mental property. However, this insistence seems little more than desperate hope. Bear in mind that if $M = B$, then $M \leftrightarrow B$ is not only actually true, but *necessarily* true (at least as a matter of nomic necessity). So even if it were true that mental properties correlate one-one with brain properties across all *actual* species, we would still be far from justified in accepting mind-brain type-identity claims. We would also need reason to believe that mental properties *had to* correlate one-one with brain properties

(Footnote 22 continued)
without comparative evaluation across species. Also see Figdor (2010) for additional critique of Bechtel and Mundale. Here Figdor also provides an informative and detailed description of the cognitive neuroscience research program that attempts to map brain structures to cognitive processes; in particular, Figdor provides a sketch of the phenomenon of degeneracy in cognitive neuroanatomy and explains how the multiple realizability claim is a live empirical hypothesis.

23 The primate's eye and the octopus eye are homoplasies—similar organs or features in different species not inherited from a common ancestor. So, according to Couch, even with homoplasies there is insufficient evidence for multiple realizability.

24 Besides Polger and Couch, those who emphasize that differences in realizers are arguably always accompanied by differences in the property realized include Shagrir (1998) and Shapiro (2008). Although, see Craver's (2004) discussion of when the "no dissociable realization" principle (that if there are two distinct realizers, then there are two distinct realized kinds) does and does not apply.

25 Relevant here is the point made by Aizawa and Gillett (2011) that despite differences between individuals in what underlies color vision (e.g., differences in photoreceptors, lens, and the macula), vision scientists still acknowledge the property of normal color vision. Here is a case in which differences in realizers do not lead scientists to deny a common realized property.

given the laws of nature that actually obtain. That there is a one-one correlation between mental properties and brain properties in every nomically possible world seems rather implausible.

So despite the actual neuroscientific practice Bechtel and Mundale mention, it remains quite doubtful that there are nomically necessary biconditionals linking mental properties to the properties of neuroscience. Still, we cannot yet be completely confident that premise 1 is true. There are two obstacles to being fully confident about premise 1.

(i) As Kim (1972) and Adams (1979) taught us (and as mentioned in 1.3), any evidence that there are significant neurological differences between the species is compatible with neural similarities alongside the differences. So any empirical evidence of neural differences in the realization of some mental property proves only that the mental property is *mildly* multiply realizable with respect to neuroscientific properties. Even if non-organic systems can have mentality, there is still no reason to deny the presence of a common neuroscientific feature since some features mentioned in neuroscience (e.g., conducting electricity) are not peculiar to organic systems. This brings us to a second obstacle to full confidence in premise 1. (ii) Even if there is no *neural* property common to all and only the bearers of some mental property, this is no reason to think the mental property is multiply realizable (in the robust sense) with respect to *all* properties of the natural sciences. There might be some chemical or physical feature of a non-neural variety that is nomically necessary and sufficient for having the mental property.[26]

In the next section I modify MR_{NI} to help us avoid these two obstacles.

1.8 Immaterial Realizers

Even if one grants that mental properties are not identical with the properties of physics, chemistry, or biology, one might still believe that every *instance* of a mental property is in fact an *instance* of some physical property or other (though not necessarily the same physical property on each occasion). While it seems this *token*-identity claim is actually correct, it also seems that the world could have been otherwise. As Lewis remarks, "[m]aterialism is meant to be a contingent thesis, a merit of our world that not all other worlds share" (1983, p. 362). Let's suppose that all mentality actually is realized purely physically. Even so, it is not implausible to think that mentality *could have been* realized in wholly *immaterial* substances—where 'immaterial' is being used here not in the physically kosher sense of being comprised of energy rather than matter, but in the more traditional sense of transcending the physical universe, as angels, spirits, and souls are supposed to do.

[26] Churchland (1982) speculated that a reductive base for cognition might be found at the thermodynamic level. And focusing on memory consolidation, Bickle (2003, Chap. 3 and 2010) argues that at the molecular level there is likely to be underlying unity.

In what sense are immaterial realizers *possible*? It's doubtful that immaterial realizers are nomically possible. By 'nomic' we are including all the laws of nature that obtain, including all laws regarding the presence of mentality. And if in fact there are no immaterial realizers, it's arguable that that's because the causal laws connecting mental properties to those of physics, chemistry, and biology prevent the occurrence of such things. However, even if immaterial realizers are nomically impossible, it does seem that they are possible in a broader sense. Immaterial realizers are certainly logically possible; nothing about the laws of logic prevents their occurrence. They also seem to be conceptually possible. While substance dualism may be false, it is not analytically false; it is not false by the very meanings of our words (although a logical behaviorist is likely to claim otherwise). Immaterial realizers would seem to be *metaphysically* possible as well. It is not just that immaterial realizers are compatible with the laws of logic (which is obviously true) and not just that they are compatible with the meanings of our mental terms (which also seems to be the case), but that immaterial realizers really could have existed. Even if there actually are no spirits or souls, there is the strong intuition that the world could have been such that these things exist.

The possibility of immaterial realizers introduces a brand of multiple realizability different from that usually discussed in the literature on psychophysical reduction. Differences in *physical* (including chemical and biological) realization are the differences usually discussed. Yet, the focus on immaterial realizers provides a more powerful argument against non-identity.

Suppose that F is multiply realizable in the following manner with respect to a class of properties *C*:

> there are members, G and H, of *C* such that G ≠ H and: (i) it is nomically possible that an instance x of G instantiates F and an instance y of H instantiates F, and (ii) it is *metaphysically* possible that there is no property I of *C* that x and y both instantiate and that is common only to instances of F.

And let's say that in this type of case F is robustly$_m$ multiply realizable with respect to *C*-properties (where 'm' stands for metaphysical).

This brand of multiple realizability is not as robust as the robust brand introduced in Sect. 1.5. Unlike that variety, this one allows that there is a *C*-property whose instantiation is nomically necessary and sufficient for the instantiation of F. Still, thanks to condition (ii), robust$_m$ multiple realizability is robust enough to entail that F is not identical to any *C*-property. For with (ii) there is no *C*-property whose presence is both metaphysically necessary and sufficient for the presence of F. Given the necessity of identity, it follows that there is no *C*-property with which F is identical.

It is true that some trivial, universally essential properties of physics, chemistry, and biology are necessary features even of immaterial substances. All possible Cartesian egos have the property of being *either a quark or not a quark* and being *comprised of ammonia molecules if comprised of ammonia molecules.* Some negative properties of the natural sciences are also essential to being an immaterial item—e.g., *not positively charged.* But, clearly, none of these trivial or negative features is *sufficient* for the instantiation of any mental property. So it seems safe to say that no property of physics, chemistry, or biology that is sufficient for some mental property is also necessary for that property—if immaterial realizers are possible. So if immaterial realizers are (metaphysically) possible, then mental properties are robustly$_m$ multiply realizable with respect to the entire class of physical properties.

Incorporating robust$_m$ multiple realizability into MR_{NI} gives us the following:

(1_m) Mental properties are multiply realizable *in the robust$_m$ sense* with respect to properties of the natural sciences.

(2_m) If mental properties are multiply realizable in the robust$_m$ sense with respect to those of the natural sciences, then there are no *metaphysically* necessary biconditionals linking the former to the latter.

(3_m) If there are no metaphysically necessary biconditionals linking mental properties to those of the natural sciences, then the former are not identical with the latter.[27]

Therefore, (4) mental properties are not identical with those of the natural sciences.

Premise 1_m seems to be true. Even if physicalism is true, even if mental phenomena do obtain solely in virtue of physical phenomena, it seems the world could have been otherwise. In particular, it seems that immaterial mental substances could have existed. And if so, then there is a metaphysically possible world in which instances of mental properties are not instances of physical properties (i.e., not instances of any physical properties which, like the trivial ones mentioned before, are indiscriminately shared by all items). Premise 2_m is certainly true. For if F is multiply realizable in the robust$_m$ sense with respect to property-set *C*, then given condition (ii) of the definition, there are no metaphysically necessary biconditionals linking F to any property of *C*. Also, given the necessity of identity, 3_m is every bit as true as premise 3 of our initial MR_{NI}.

The merit of this latest version of MR_{NI} is that it offers a clear way to avoid the two obstacles to proving non-identity noted at the close of 1.7. Recall: (i) there is the difficulty proving that mental properties are robustly multiply realizable, in our original nomic sense, with respect to neuroscientific properties, since any empirical evidence of neural differences in the realization of a mental property is consistent with the mental property being only *mildly* multiply realizable (allowing that there are neural similarities despite the differences), and (ii) there is the difficulty showing that mental properties are robustly multiply realizable, in the nomic sense, with respect to *all* properties of the natural sciences, and not just neuroscientific properties. With our new first premise, 1_m, both obstacles are avoided. Suppose there

[27] As Melnyk (2008, p. 1294) mentions in a footnote, a "genuine multiple *realizability* objection would appeal to the metaphysical *possibility* of, say, pain without the neurophysiological state thought to be identical with pain; such a bare possibility would be enough to contradict the proposed identity claim, given the necessity of identity."

were a neural feature that corresponds one-one with some mental property and does so as a matter of nomic necessity. Still, with the metaphysical possibility of immaterial realizers, this correspondence between the mental and neural property fails to obtain in some possible worlds. Likewise, any one-one correlation between a mental property and a property of any other natural science would also fail to obtain in some possible worlds. Given the necessity of identity, it follows that the mental property is not identical with any property of the neurosciences or any other natural science.

One might feel uncomfortable with the talk of *metaphysical* possibility/necessity. One might think that while it makes sense to speak of logical, nomological, or physical possibility/necessity, it makes no sense to speak of possibility/necessity *simpliciter*, which is what those who speak of the metaphysical brand of modality have in mind. For anyone who wishes to avoid talk of metaphysical possibility/necessity, I offer one final version of MR_{NI}.

Recall Lewis' claim that "[m]aterialism is meant to be a contingent thesis, a merit of our world that not all other worlds share." Perhaps materialism (physicalism) is true of the actual world. Even so, it seems that the world could have been otherwise. With Lewis, Terence Horgan supports the *actual* truth of physicalism while acknowledging that the world *could have* contained immaterial items—ghosts, spirits, etc. Moreover, the world might have contained these spooky items, Horgan contends, even with all of the physical laws that actually obtain:

> [w]e who claim that the microphysical facts determine all the facts want to deny that such beings [Cartesian souls, angels or gods] exist in our actual world. However, we need not deny that there are some possible worlds in which they do exist and in which the microphysical laws of our world are never violated. In such worlds the spirits would not interfere with the ordinary operations of physical laws upon physical substances; they would simply co-exist with the physical (1982, pp. 34–35).

Provided that the immaterial items do not causally interfere with the physical world, their existence is consistent with the actual physical laws. Immaterial items could have existed, even with all of the actual physical laws in place, provided that the spooky extras do not causally interfere with the way the world is physically. If Horgan is right about this, and it seems that he is, then immaterial realizers are *physically* possible.

So let us consider our final variety of robust multiple realizability. Suppose that F is multiply realizable with respect to *C* in the following manner:

> there are members, G and H, of *C* such that G ≠ H and: (i) it is nomically possible that an instance x of G instantiates F and an instance y of H instantiates F, and (ii) it is *physically* possible that there is no property I of *C* that x and y both instantiate and that is common only to instances of F.

And let's say in this type of case that F is robustly_p multiply realizable with respect to *C*-properties. With the notion of robust_p multiple realizability, we can modify MR_{NI} one last time to read:

(1_p) Mental properties are *robustly$_p$* multiply realizable with respect to properties of the natural sciences.

(2_p) If mental properties are robustly$_p$ multiply realizable with respect to those of the natural sciences, then there are no *physically* necessary biconditionals linking the former to the latter.

(3_p) If there are no physically necessary biconditionals linking mental properties to those of the natural sciences, then the former are not identical with the latter.

Therefore, (4) mental properties are not identical with those of the natural sciences.

Premise 1_p appears to be true. Even if we are physicalists, even if we think mental phenomena obtain solely in virtue of physical phenomena, it seems there's no reason to deny that the world could have been otherwise. It seems that some immaterial items could have existed, and as Horgan notes, they could have existed even with all the actual physical laws in place—provided that the immaterial items do not interfere with the operations of the physical world. If so, then in some physically possible world, mental property M is instantiated, but there is no physical property P, such that all instances of M are instances of P.[28] And in that case, M is multiply realizable in the robust$_p$ manner with respect to the set of physical properties.

Premise 2_p is certainly true. If mental properties are robustly$_p$ multiply realizable, then given condition (ii) in the definition, there are no physically necessary biconditionals linking the mental property to some physical property. Also, given the necessity of identity, 3_p is just as true as 3 of our initial MR_{NI}.

As with the metaphysically infused version of MR_{NI}, the two obstacles to accepting the first premise are avoided with this version as well. Suppose there were a neural feature that corresponds one-one with some mental property and does so as a matter of nomic necessity. Still, with the physical possibility of immaterial realizers, this correspondence between the mental and neural property does not obtain in every physically possible world. So given the necessity of identity, the mental property is not identical with any neural property. Likewise, given the physical possibility of immaterial realizers, any actual or nomically necessary correlation between the mental property and a property of any other natural science would also fail to obtain in some physically possible worlds. So, given the necessity of identity, the mental property is not identical with any property of the natural sciences.

There is, however, one more obstacle we need to address before we can confidently accept this latest version (or any of our versions) of MR_{NI}. This obstacle is a notorious objection to multiple realizability arguments that we have not yet addressed.

---

[28] We're imagining again that P is some non-trivial physical property—unlike, e.g., either a quark or not a quark, indiscriminately had by all items.

1.9 The Disjunctive Strategy

Suppose that P is the disjunction of all nomically possible realizers of some mental property M. Also suppose that each disjunct of P is nomically sufficient for the presence of M; i.e., given the laws of nature, the instantiation of the disjunct ensures the instantiation of M.[29] Suppose, further, that each disjunct of P is a property of one of the natural sciences (e.g., a property of physics, chemistry, or biology). Then there is a nomically necessary biconditional linking M to a property, P, of the natural sciences. So, it seems, while M is not nomically equivalent to any non-disjunctive property of the natural sciences, there is a property of the natural sciences with which M is nomically equivalent, and therefore with which M might qualify as identical.

Fodor rejects the disjunctive strategy. He claims that "a necessary condition on a universal generalization being lawlike is that the predicates which constitute its antecedent and consequent should pick out natural kinds" (1974, p. 108). Yet, it seems that for any disjunctive property nomically equivalent to some mental property, the various disjuncts will be widely heterogeneous (including not only all the different possible neural realizers, but also all the possible realizers of the mental property in inorganic systems). So, Fodor concludes, any disjunctive property nomically equivalent to some mental property will not count as a natural kind.

The idea that the properties employed by the disjunctive strategy are excessively heterogeneous is a common complaint.[30] Actually, there are at least two distinct heterogeneity complaints that might be made: that (i) heterogeneous disjunctions do not denote properties belonging to the natural sciences, and that (ii) heterogeneous disjunctions do not denote genuine properties of any sort. However, both (i) and (ii) can plausibly be resisted. Perhaps there is good reason (concerning explanatory ability or predictive power) to deny that disjunctive properties with heterogeneous disjuncts are the sort to figure in the sciences. But even if this is correct, which is controversial,[31] it does not tell against their being genuine properties, at least not without the contentious view that science names the only properties that exist. Even if we endorse a causal account of properties, and one which entails that all instances of the same property share the causal powers distinctive of that property, one need not and arguably should not deny that disjunctive predicates denote genuine properties. For it is arguable that the diverse disjuncts are causally similar at least

---

[29] To be nomically sufficient for the mental property, each disjunct will be what Shoemaker (2007, p. 21) calls a "total" realizer and not merely a "core" realizer. The core realizer of a mental property is what plays the definitive functional role, whereas the total realizer includes not just the core realizer, but whatever relations the core realizer bears to other states of the organism to ensure that it plays the requisite functional role. So each disjunct will be a highly complex physical property.

[30] In addition to Fodor, critics of the disjunctive strategy include Pereboom and Kornblith (1991), Zangwill (1995), and Seager (1991).

[31] See Clapp's (2001) and Walter's (2006) reasons for thinking that disjunctive properties with heterogeneous disjuncts can qualify as properties of the sciences.

insofar as they are able to play the functional role definitive of the mental property.[32] So the attempt to reject the disjunctive strategy on the basis of either (i) or (ii) is not without controversy.

The appeal to immaterial realizers, however, avoids the controversy of (i) and (ii), providing a more effective response to the disjunctive strategy. Suppose that P is a disjunctive property, whose disjuncts are realizers of mental property M, each of which is sufficient for the presence of M. Now, for it to be true that M = P, the disjuncts of P will need to include all nomically possible realizers of M.[33] And all nomically possible realizers of M include all physically possible realizers of M. So if it is physically possible for M to have immaterial realizers, then for it to be true that M = P, some of the disjuncts of P will not be properties of physics, chemistry, neurobiology, or any other natural science. Thus, given the physical possibility of immaterial realizers, M is identical with a disjunctive property only if that disjunctive property contains disjuncts that are not properties of the natural sciences. It would appear, then, that the physical possibility of immaterial realizers neatly undermines the disjunctive strategy of reducing mental properties to those of the natural sciences.[34]

1.10 Concluding Remarks

As Bechtel and Mundale show, the empirical evidence and facts about neuroscientific methodology suggest that mental properties are not likely to be as multiply realizable as philosophers are inclined to think. However, given the actual evidence for neural differences across the species as well as neural variation within the life of the same individual, it still seems doubtful that mental properties correspond one-to-one with neural properties across all actual species of thinking and feeling organisms. And yet, despite the reason for doubt, there are two obstacles to confidently concluding non-identity. (i) Any empirical evidence of neural differences in the realization of some mental property M proves only that M is *mildly* multiply realizable with respect to neuroscientific properties, for even with great neural differences in the instantiation of M, there may still be crucial neural similarities.

[32] And this similarity in causal role would seem to require some similarity in physical structure, as Walter indicates (2006, p. 58).

[33] Given the necessity of identity, all metaphysically possible realizers will also need to be included.

[34] One who is already convinced that mental properties are identical with physical, chemical, or biological properties might rely on the necessity of identity and simply deny the possibility of immaterial realizers. However, this is not a viable response for proponents of the disjunctive strategy, since considerations of multiple realizability have already led them to reject the identity of mental properties with (non-disjunctive) properties of the natural sciences. Without the belief in these non-disjunctive type-identities, there is no good reason for them to reject the physical possibility of immaterial realizers.

The neural similarities might extend even to non-organic systems since not all neuroscientific features are peculiar to organic systems. Moreover, (ii) even assuming there is no neural property common to all and only the bearers of M, this is no reason to think that M is robustly multiply realizable with respect to properties of *each of the natural sciences*.

The appeal to either the metaphysical possibility or the physical possibility of immaterial realizers helps us avoid these two obstacles. For even if each mental property actually did correspond one-to-one with some physical, chemical, or biological property, and even corresponded as a matter of nomic necessity, the physical (not to mention metaphysical) possibility of immaterial realizers shows that mental properties are not identical with any neuroscientific properties, and also not identical with any properties of any of the other natural sciences. In addition, the physical possibility of immaterial realizers undermines a third obstacle to denying type-identities. The disjunctive strategy appeals to the one-to-one correspondence between some mental property M and the disjunction of each of the nomically possible realizers of M. Yet, with the physical possibility of immaterial realizers, some of the disjuncts of the disjunctive property with which M corresponds as a matter of physical necessity are not properties of the natural sciences. That's enough, given the necessity of identity, to entail non-identity.

Suppose that at least one of our versions of MR_{NI} is sound and therefore that the conclusion is true—mental properties are not identical with properties of the natural sciences. The truth of this conclusion would, at least initially, seem to be at odds with the doctrine of *materialism*, more commonly called "*physicalism*" in recent decades.[35] The physicalist regarding mentality believes that mental phenomena obtain *solely in virtue of physical phenomena*. However, if mental properties are not identical with properties of physics or any of the other natural sciences, then it is not clear how this view could be true. For if mental properties are not identical with properties of physics or any of the other natural sciences, then it seems that mental phenomena are not due solely to physical phenomena.

Yet, many philosophers who accept the conclusion of multiple realizability arguments also claim to be physicalists. For those who support *non-reductive physicalism*, there are two popular ways to explain how it can be that mentality depends entirely on physical phenomena despite the lack of true psychophysical type-identity claims. There is the highly popular appeal to the notion of *supervenience*. There also is the idea that mental properties, while not physical, are *realized* physically. Despite the popularity of these two appeals, in the next two chapters I argue that neither succeeds in showing how physicalism might be true without mental properties being identical with physical properties. The answer to how one can accept the conclusion of the multiple realizability arguments while still earning the label 'physicalist' is revealed in the final chapter.

---

[35] A main reason that 'physicalism' came to be considered a more appropriate label is that the physical sciences are not restricted to what qualify as *material* items.

References

Aizawa, K. 2009. Neuroscience and multiple realization: A reply to Bechtel and Mundale. *Synthese* 167: 493–510.
Aizawa, K., and C. Gillett. 2009. The (multiple) realization of psychological and other properties in the sciences. *Mind and Language* 24: 181–208.
Aizawa, K., and C. Gillett. 2011. The autonomy of psychology in the age of neuroscience. In *Causality in the sciences*, ed. P.M. Illari, F. Russo, and J. Williamson, 202–223. Oxford: Oxford University Press.
Armstrong, D. 1968. *A materialist theory of the mind*. London: Routledge.
Adams, F. 1979. Properties, functionalism, and the identity theory. *Eidos* 1: 153–179.
Bechtel, W., and J. Mundale. 1999. Multiple realizability revisited: Linking cognitive and neural states. *Philosophy of Science* 66: 175–207.
Bickle, J. 1992. Multiple realizability and psychophysical reduction. *Behavior and Philosophy* 20: 47–58.
Bickle, J. 1998. *Psychoneural reduction: The new wave*. Cambridge, MA: MIT Press.
Bickle, J. 2003. *Philosophy and neuroscience: A ruthlessly reductive account*. Dordrecht: Kluwer.
Bickle, J. 2010. Has the last decade of challenges to the multiple realization argument provided aid and comfort to psychoneural reductionists? *Synthese* 177: 247–260.
Block, N. 1980. Introduction: What is functionalism? In *Readings in philosophy of psychology*, vol. I, ed. N. Block, 171–184. Cambridge, MA: Harvard University Press.
Block, N. 1981. Psychologism and behaviorism. *Philosophical Review* 90: 5–43.
Buonomano, D., and M. Merzenich. 1998. Cortical plasticity: From synapses to maps. *Annual Review of Neuroscience* 21: 149–186.
Carnap, R. 1932/1933. Psychology in physical language. *Erkenntnis* 3: 107–142.
Churchland, P.M. 1979. *Scientific realism and the plasticity of mind*. Cambridge: Cambridge University Press.
Churchland, P.M. 1982. Is 'thinker' a natural kind? *Dialogue* 21: 223–238.
Churchland, P.M. 1984. *Matter and consciousness*. Cambridge, MA: The MIT Press.
Churchland, P.M. 1985. Reduction, qualia, and the direct introspection of brain states. *The Journal of Philosophy* 82: 8–28.
Clapp, L. 2001. Disjunctive properties: Multiple realizations. *The Journal of Philosophy* 98: 111–136.
Couch, M. 2004. Discussion: A defense of Bechtel and Mundale. *Philosophy of Science* 71: 198–204.
Craver, C. 2004. Dissociable realization and kind splitting. *Philosophy of Science* 71: 960–971.
Draganski, B., C. Gaser, G. Kempermann, H.G. Kuhn, J. Winkler, C. Büchel, and A. May. 2006. Temporal and spatial dynamics of brain structure changes during extensive learning. *The Journal of Neuroscience* 26: 6314–6317.
Enc, B. 1983. In defense of the identity theory. *The Journal of Philosophy* 80: 279–298.
Endicott, R. 1989. On physical multiple realization. *Pacific Philosophical Quarterly* 70: 212–224.
Endicott, R. 1993. Species-specific properties and more narrow reductive strategies. *Erkenntnis* 38: 303–321.
Endicott, R. 2005. Multiple realizability. In *Encyclopedia of philosophy*, 2nd ed, ed. D. Borchert, 427–432. Detroit: Thomson Gale/Macmillan Reference.
Figdor, C. 2010. Neuroscience and the multiple realization of cognitive functions. *Philosophy of Science* 77: 419–456.
Fodor, J. 1974. Special sciences (or: the disunity of science as a working hypothesis). *Synthese* 28: 97–115.
Francescotti, R. 1997. What multiple realizability does not show. *Journal of Mind and Behavior* 18: 13–28.
Funkhouser, E. 2007. A liberal conception of multiple realizability. *Philosophical Studies* 132: 467–494.

Gillett, C. 2003. The metaphysics of realization, multiple realizability, and the special sciences. *The Journal of Philosophy* 100: 591–603.
Hempel, C. 1949. The logical analysis of psychology. In *Readings in philosophical analysis*, ed. H. Feigl, and W. Sellars, 373–384. New York: Appleton-Century-Crofts.
Hill, C. 1991. *Sensations: A defense of type materialism*. Cambridge: Cambridge University Press.
Hooker, C. 1981. Towards a general theory of reduction. Part I: Historical and scientific setting. Part II: Identity in reduction. Part III: Cross-categorial reduction. *Dialogue* 20:38–59, 201–236, 496–529.
Horgan, T. 1982. Supervenience and microphysics. *Pacific Philosophical Quarterly* 63: 29–43.
Horgan, T. 1993. Nonreductive materialism and the explanatory autonomy of psychology. In *Naturalism: A critical appraisal*, ed. S.J. Wagner, and R. Warner, 295–320. South Bend, IN: University of Notre Dame Press.
Kim, J. 1972. Phenomenal properties, psychophysical laws, and the identity theory. *Monist* 56: 177–192.
Kim, J. 1989. The myth of nonreductive materialism. *Proceedings and Addresses of the American Philosophical Association* 63: 31–47.
Kim, S. 2002. Testing multiple realizability: A discussion of Bechtel and Mundale. *Philosophy of Science* 69: 606–610.
Lewis, D. 1966. An argument for the identity theory. *The Journal of Philosophy* 63: 17–25.
Lewis, D. 1969. Review of *Art, mind, and religion*. *The Journal of Philosophy* 66: 22–27.
Lewis, D. 1980. Mad pain and Martian pain. In *Readings in philosophy of psychology*, vol. I, ed. N. Block, 216–222. Cambridge, Mass: Harvard University Press.
Lewis, D. 1983. New work for a theory of universals. *Australasian Journal of Philosophy* 61: 343–377.
Melnyk, A. 2008. Can physicalism be non-reductive? *Philosophy Compass* 3: 1281–1296.
Nagel, E. 1961. *The structure of science: Problems in the logic of scientific explanation*. New York: Harcourt, Brace, and World.
Pereboom, D., and H. Kornblith. 1991. The metaphysics of irreducibility. *Philosophical Studies* 63: 125–145.
Place, U.T. 1956. Is consciousness a brain process? *British Journal of Psychology* 47: 44–50.
Polger, T. 2004. *Natural minds*. Cambridge, Mass.: MIT Press.
Polger, T. 2007. Realization and the metaphysics of mind. *Australasian Journal of Philosophy* 85: 233–259.
Polger, T. 2008. Two confusions concerning multiple realization. *Philosophy of Science* 75: 537–547.
Polger, T. 2009. Evaluating the evidence for multiple realization. *Synthese* 167: 457–472.
Polger, T., and L. Shapiro. 2008. Understanding the dimensions of realization. *The Journal of Philosophy* 105: 213–222.
Putnam, H. 1965. Brains and behavior. In *Analytical philosophy: Second series*, ed. R. Butler, 1–19. Oxford: Blackwell.
Putnam, H. 1967. Psychological predicates. In *Art, mind, and religion*, ed. W.H. Capitan, and D.D. Merrill, 37–48. Pittsburgh: University of Pittsburgh Press.
Richardson, R. 1979. Functionalism and reductionism. *Philosophy of Science* 46: 533–558.
Richardson, R. 1982. How not to reduce a functional psychology. *Philosophy of science* 49: 125–137.
Richardson, R. 2008. Autonomy and multiple realization. *Philosophy of Science* 75: 526–536.
Richardson, R. 2009. Multiple realization and methodological pluralism. *Synthese* 167: 473–492.
Ryle, G. 1949. *The concept of mind*. London: Hutchinson.
Seager, W. 1991. Disjunctive laws and supervenience. *Analysis* 51: 93–98.
Searle, J. 1992. *The rediscovery of the mind*. Cambridge, Mass.: MIT Press.
Shagrir, O. 1998. Multiple realization, computation and the taxonomy of psychological states. *Synthese* 114: 445–461.
Shapiro, L. 2000. Multiple realizations. *The Journal of Philosophy* 97: 635–654.
Shapiro, L. 2008. How to test for multiple realization. *Philosophy of Science* 75: 514–525.

Shoemaker, S. 2007. *Physical Realization*. New York: Oxford University Press.
Smart, J.J.C. 1959. Sensations and brain processes. *The Philosophical Review* 68: 141–156.
Stein, D., and S. Hoffman. 2003. Concepts of CNS plasticity in the context of brain damage and repair. *The Journal of Head Trauma Rehabilitation* 18: 317–341.
Walter, S. 2006. Multiple realizability and reduction: A defense of the disjunctive move. *Metaphysica* 7: 43–65.
Wieloch, T., and K. Nikolich. 2006. Mechanisms of neural plasticity following brain injury. *Current Opinion in Neurobiology* 16: 258–264.
Zangwill, N. 1995. Supervenience, reduction, and infinite disjunction. *Philosophia* 24: 321–330.

Chapter 2
Supervenience and Physicalism

Thanks to the lure of functionalist accounts of mentality and the wide degree of multiple realizability they entail, a dominant view in the philosophy of mind for the past few decades is the belief that mental properties are not identical with neural properties or any other properties of the natural sciences. Yet, many of those who endorse this non-reductionist view also claim to be physicalists, believing that all concrete phenomena obtain *solely in virtue of, depending on nothing other than*, physical phenomena. This pair of views (often called "non-reductive physicalism") is at least initially surprising since it is not entirely clear how mental phenomena can obtain solely in virtue of physical phenomena if mental properties are not physical properties. If mental properties are not physical, then it would seem that mental phenomena (which obviously involve the instantiation of mental properties) are something more or other than physical phenomena, in which case, the former do not depend wholly on the latter. So non-reductive physicalists have some explaining to do. The task is to explain how one can consistently deny that mental properties are physical while remaining faithful to physicalism.

One popular way to make sense of how physicalism might be true even if mental properties are not physical is by appealing to the notion of *supervenience*. The notion of supervenience offers hope for understanding how it can be that a class of properties, A, is distinct from a class of properties, B, even while A-properties are a function of nothing other than B-properties. I do not object to relying on the notion of supervenience in a definition of physicalism; as explained in Sects. 2.2 and 2.3, it seems that the right brand of supervenience along with a robust physical composition requirement can capture all that physicalism demands. What I object to, instead, is the appeal to supervenience as a way to make sense of how one can be a physicalist while denying that mental properties are physical. As shown here, rather than revealing how a non-reductionist can remain faithful to physicalism, the notion of supervenience only serves to make the *prima facie* tension between physicalism and non-reductionism even more apparent. For, as described in Sect. 2.4, it is a mystery why the supervenience required by physicalism should be thought to obtain if mental properties are not physical.

There are numerous ways to formulate the idea that mental properties supervene on physical properties, and the philosophical literature is replete with descriptions of

R. Francescotti, *Physicalism and the Mind*,
SpringerBriefs in Philosophy, DOI 10.1007/978-94-017-9451-0_2

various brands of supervenience. Despite the vast discussion on the subject, I shall review only enough of the details regarding supervenience needed to appreciate what I consider the main threat to non-reductive physicalism, described in Sect. 2.4.

2.1 Supervenience: Background

The basic idea is that a class of properties A supervenes on a class of properties B just in case a difference in A-properties requires some difference in B-properties. As Lewis crisply puts, "no difference of one sort without differences of another sort" (1986, p. 14). Suppose that no two individuals and no two actions could differ morally without differing in some physical respect. Then moral properties supervene on physical properties. On the other hand, if there could be a moral difference in the absence of any physical difference, then moral properties fail to supervene on physical properties. Likewise, to say that mental properties supervene on physical properties is to say, as Davidson puts it, "that there cannot be two events alike in all physical respects but differing in some mental respect, or that an object cannot alter in some mental respect without altering in some physical respect" (1970, p. 88).[1,2]

Here Davidson describes a *local* supervenience thesis. The claim is that no two *individuals* (objects or events) can differ mentally without differing physically. One might instead opt for a *global* supervenience thesis, which is what Haugeland expresses when he writes, "[t]he world could not have been different in any respect, without having been different in some strictly physical respect" (1984, p. 1). Here the claim that non-physical differences require physical differences is applied to possible worlds as a whole. Applied to mentality, the global thesis is that the world could not be different in any mental respect without differing physically.

If mental (or moral) differences require physical differences, then physical sameness entails mental (moral) sameness. Thus, as Kim expresses the notion of global supervenience, for any sets of properties, A and B, A globally supervenes on B "just in case worlds that are indiscernible with respect to B ("B-indiscernible," for short) are also A-indiscernible" (1984, p. 168). When applied to the mental and the physical, the claim is that

(GS) for any possible worlds, w_1 and w_2, if w_1 and w_2 are physically indiscernible, then w_1 and w_2 are mentally indiscernible.

---

[1] See Kim (1990) for a nice review of the history of the term 'supervenience', from its use in the moral philosophy of Henry Sidgwick, G.E. Moore, and R.M. Hare, to the writings of the British emergentists, and to its use by Davidson (1970) which sparked renewed interest in the concept of supervenience in contemporary philosophy of mind.

[2] From here on talk of properties is meant to include *relational* properties. So the claim is that items differing in their mental properties, including mental *relations*, differ in their physical properties/relations.

Of course, *worlds* themselves don't have mental properties. Individuals in worlds have the mental properties. So to say that worlds w_1 and w_2 are mentally (or physically) the same is to say that the distribution of mental (or physical) properties over individuals is the same in both worlds.

The *local* counterpart to GS is:

(LS) for any possible worlds, w_1 and w_2, and any items, x in w_1 and y in w_2, if x in w_1 is physically indiscernible from y in w_2, then x in w_1 is mentally indiscernible from y in w_2.

Or instead of LS one might choose as a local supervenience thesis what Kim (1984) famously calls "strong" supervenience. To say that mental properties strongly supervene on physical properties is to say that

(SS) necessarily, for any mental property M, and any item x that has M, there is a physical property, or set of physical properties, P, such that x has P, and necessarily, for any item y, if y has P, then y has M.

The second occurrence of 'necessarily' distinguishes SS from what Kim (1984) calls "weak" supervenience. With the second modal operator SS (like LS) guarantees that physical duplicates are mental duplicates *across* as well as within possible worlds, capturing the idea that individuals *could not* differ mentally without differing physically.[3,4]

Whether any of these accurately captures an essential component of physicalism depends on whether physicalism is a *contingent* thesis. It is arguable that even if physicalism is actually true, it could have been false. Recall, again, Lewis' claim that "[m]aterialism is meant to be a contingent thesis, a merit of our world that not all other worlds share" (1983, p. 362). If physicalism is contingent, then to capture

---

[3] McLaughlin (1995, Sect. 2) calls SS the *modal operator* version of strong supervenience and LS the *possible worlds* version. And as McLaughlin notes, there is a significant difference between SS and LS. SS requires that an individual has mental properties only by having physical properties, but LS allows that one can have mental properties without any physical properties. So with LS there could be a mental duplicate of you or me who lacks physical properties altogether. However, this result can be avoided if we include in the set of subvening properties the *negation* of certain properties we normally consider physical. For example, if *not having weight* and *not instantiated in physical space* are regarded as physical properties, then the possibility of mental items being wholly non-physical is avoided. So a version of LS may be rendered logically equivalent to a version of SS by making the supervenience base closed under complementation, where for each property in the base set its negation is also included.

[4] One might wish to avoid the reference to individuals in LS and SS and talk about the *regions of space* they occupy. With his notion of a *P-region*, a spatio-temporal region of a physically possible world, Horgan (1993a, p. 571) expresses his *regional supervenience* thesis as: "There are no two *P*-regions that are exactly alike in all qualitative intrinsic physical features but different in some other qualitative intrinsic feature." Also see Horgan (1982, p. 37).

its content, the supervenience claims above will need rephrasing. To allow worlds where the physical facts do not fix the mental facts, GS might be replaced with

(GS*) for any possible world w, if w is physically indiscernible from the actual world, then w and the actual world are mentally indiscernible.

LS and SS can also be modified to render physicalism contingent. Like GS, LS may be changed to quantify over worlds that are physical duplicates of the actual world, and the *first* modal operator in SS may be dropped.

The debate over which brand of supervenience thesis is best fit to characterize physicalism is surprisingly large and complex. There is the issue, just mentioned, of whether to endorse a supervenience thesis that renders physicalism contingent. There is also the major issue of whether to prefer a global supervenience thesis or a local one.[5] And there is vast discussion of whether a global supervenience thesis may be considered logically equivalent to a local thesis, and which types of properties need to be allowed in the supervenience base (including which property-forming closure principles the base set would need to satisfy) to make a global thesis equivalent to a local one.[6]

Another widely discussed issue is whether we should characterize physicalism in a way that allows the possibility of mental *extras*. The idea that the physical facts fix the mental facts entails that a world that physically duplicates the actual world has all the actual mental episodes, but this seems to allow that the world might also have some additional mental extras. Provided that these mental additions do not interfere with any of the physical laws that obtain, the physicalist view that all actual mentality is solely a function of the physical seems to be preserved. Recall Horgan's point (quoted in 1.8) that while physicalists insist that immaterial mental substances do not in fact exist, they "need not deny that there are some possible

[5] Relevant here is Kim's (1987, pp. 322–324) point that a local supervenience thesis can apply to *extrinsic* supervening properties by allowing extrinsic properties in the base set. Also, by employing Kim's (1988) notion of "multiple domain" supervenience, we can allow with a local thesis that the individuals with the subvening properties differ from those that have the supervening properties (e.g., with the bearers of the physical properties being proper parts of those that have the mental properties). There's also Kim's (1987, p. 321) famous distant ammonia molecule example which suggests that global supervenience claims allow mental differences despite only seemingly irrelevant physical differences.

[6] In an oft-cited debate, Petrie (1987) argues that contrary to what Kim (1984, p. 168) suggests, GS does not entail SS. Paull and Sider (1992) point out that Petrie's method of proving the inequivalence is inadequate and they provide an improved example to show the inequivalence, while also noting that when restricted to *intrinsic* properties, the global and strong theses are equivalent. Kim, who initially (1987, p. 318) concedes Petrie's point about the inequivalence, later claims that "if extrinsic properties are included in both the supervenient and subvenient sets—in particular, if, along with the usual Boolean operations, identity and quantification are allowed for property composition—again the equivalence will obtain" (1993, p. 170). Stalnaker (1996, p. 238) provides the proof of the equivalence. Also see Moyer's (2008) appeal to an intuitively plausible "Recombination Principle" to argue that even *weak* (local) supervenience is equivalent to SS.

worlds in which they do exist and in which the microphysical laws of our world are never violated." In those possible worlds, "the spirits would not interfere with the ordinary operations of physical laws upon physical substances; they would simply co-exist with the physical" (1982, pp. 34–35). If the possibility of mental extras is consistent with physicalism, then our supervenience thesis used to define physicalism should allow that some physical duplicates of the actual world are not mentally duplicates (since they contain mental extras). For this reason, Jackson (1994, p. 28) introduces the idea of a *minimal* physical duplicate of the actual world, which is what you get if you duplicate the actual world physically *and stop right there*. With the notion of a minimal physical duplicate, Jackson characterizes physicalism as the view that all minimal physical duplicates of the actual world are mental duplicates of the actual world. This allows that there are non-minimal physical duplicates with mental extras, which arguably is consistent with physicalism. Or we might choose Chalmers' (1996, p. 40) method of allowing for extras —by construing physicalism as the claim that any physical duplicate of our world has all the *positive* facts of our world. This allows that physical duplicates of the actual world differ from the actual world by having additional positive facts regarding the presence of immaterial extras (and thereby lacking actual negative facts such as the fact that there are no immaterial mental items).[7]

Another source of debate is that it's not entirely clear how to understand talk of the *same distribution* of properties in global supervenience theses. For one thing, it is not clear whether the same distribution should be considered a distribution over the *same objects*. Consider, for instance, those who deny transworld identity by endorsing counterpart theory. These philosophers will not regard non-actual worlds that physically duplicate, and therefore mentally duplicate, the way things actually are as worlds with the same items as those that actually exist. And if we allow that the worlds being compared do not contain the same items, then questions arise regarding which distributions of subvening properties over individuals are supposed to ensure which distributions of supervening properties over individuals. For these reasons, we can expect different possible ways to construe a global supervenience thesis. For example, Stalnaker (1996) and McLaughlin (1997) distinguish between *strong* and *weak* global supervenience. Consider any two possible worlds, w_1 and w_2, and a one-one mapping of each item x in w_1 onto some item y in w_2, and suppose that this isomorphism is ϕ-preserving for some class of properties, ϕ, just in case for every property F in ϕ, Fx in w_1 just in case Fy in w_2. Then we can say that a class of properties, A, *strongly* globally supervenes on a class of properties, B, just in case for any w_1 and w_2, *every* B-preserving isomorphism between w_1 and w_2 is an A-preserving isomorphism between them. And A-properties *weakly* globally supervene on B-properties just in case for any w_1 and w_2, if there is a B-preserving isomorphism between w_1 and w_2, then there is an A-preserving isomorphism

[7] For other ways to restrict a supervenience thesis to allow for immaterial extras, see Horgan (1982, p. 36), Lewis (1983, p. 364), Leuenberger (2008), and Yoshimi (2007). (Although, in Francescotti (2014) I argue that physicalists should avoid these restricted supervenience claims and not allow mental extras).

between them. Bennett (2004) adds a middling variety of global supervenience, which obtains just in case for any w_1 and w_2, if there is a B-preserving isomorphism between them, then at least one B-preserving isomorphism between them is itself an A-preserving isomorphism. For discussion of which of these brands of global supervenience, if any, best captures our physicalist intuitions, see Bennett (2004), Leuenberger (2009), and Shagrir (2009, 2013).

In what follows, I shall ignore each of the technical issues mentioned above, and that's because none are crucial to the main focus of this chapter, which is the issue of whether a supervenience thesis enables one to earn the title 'physicalist' if one denies that mental properties are physical properties. A choice to be made among the various brands of supervenience that is crucial to this issue is presented in Sect. 2.3. However, before getting to that important choice, it should first be noted that while a supervenience thesis may be a necessary component of physicalism, there is one major reason that it cannot satisfy our physicalist intuitions all by itself. The reason I have in mind is the widely recognized point that the supervenience of the mental on the physical is perfectly compatible with mental properties being instantiated in a wholly non-physical fashion.

2.2 Supervenience and Substance Dualism

While supervenience claims tell us how mental properties covary with physical properties, they do not tell us how mental properties are realized (i.e., instantiated). Thus, a supervenience claim in itself leaves open the possibility that mental properties are realized in ways wholly unacceptable by physicalist standards. In fact, a supervenience thesis is compatible not only with property dualism (as the non-reductive physicalist would hope[8]) but also with *substance dualism*. Suppose that any bearer of mental properties is comprised of an immaterial soul, existing in some non-physical realm, in addition to a physical body. Suppose also that all immaterial souls are dependent on the operations of physical bodies in such a way that any variation in soul properties occurs only with a variation in physical properties of the body. Then any physical duplicates of this world will be mental duplicates of this world. So this scenario is compatible with varieties of global supervenience. This scenario is compatible even with Kim's strong supervenience. Suppose that mental properties are instantiated only in immaterial souls. Suppose, also, that necessarily any individual x that has an immaterial soul has a physical body on which that soul is dependent, and dependent in such a way that each mental property of the soul is necessitated by some physical property of the body. In this case, necessarily, for any mental property M of individual x, there is some physical property P, such that necessarily for any individual y, if y has P, then y has

[8] By denying that mental properties are physical properties, the non-reductionist does qualify as a property dualist even if there is an allegiance to physicalism.

M. So the mental strongly supervenes on the physical in this case despite the substance dualism.[9]

Whatever else physicalism allows, it certainly does not allow that substance dualism is true. Clearly, then, the content of physicalism is not adequately captured with a supervenience thesis alone. However, one might still think there is hope for an adequate formulation of physicalism that is largely supervenience-based. To avoid substance dualism, we simply need to conjoin a supervenience thesis with some constraint on the *composition* of mental items. The constraint might be what Hellman and Thompson (1975) called the "Principle of Physical Exhaustion," according to which, "everything concrete is *exhausted* by basic physical objects" (p. 555).[10] Applied to mentality, this principle amounts to the following *physical decomposition* claim,

(PD) each mental particular (i.e., each instance of a mental property or bearer of mental properties) is either a physical item or possesses a decomposition into parts, all of which are physical items.

PD alone is not enough to satisfy our physicalist intuitions. Since it places no constraints on how mental properties covary with physical properties, PD is compatible with there being a possible world that is physically indistinguishable from the actual world but with a radically different distribution of mental properties, and perhaps completely devoid of mentality. At the same time, a supervenience thesis does not itself fully capture the content of physicalism, for it allows that mental processes are purely immaterial. However, by conjoining the two (supervenience plus PD), one might suspect that we do express all that physicalism requires.[11]

Whether the combo of PD and a supervenience thesis really does provide sufficient conditions for the truth of physicalism depends on the answer to another major issue regarding the supervenience relation. Supervenience theses, whether in

---

[9] I am imagining that the x and y are thinkers or feelings that bear the mental properties. But the individual items that have mental properties also include particular mental events/episodes; an episode that instantiates mental property M has the mental property of *being an instance of M*. Despite the substance dualism, strong supervenience can hold for mental episodes as well. For any instance of mental property M that occurs within one of these immaterial souls, there is the (apparently) physical property, *being dependent on P*, where having this physical property necessitates being an instance of M.

[10] Or, as Pettit puts it, "[e]verything in the empirical world is composed in some way—composed without remainder—out of (subatomic) entities of the kind that microphysics posits, or it is itself uncomposed and microphysical" (1993, p. 215).

[11] Hellman and Thompson add to their Principle of Physical Exhaustion a commitment to *truth-determination*: "If one kind or realm of facts determines another, then, at a minimum, the truth values of sentences expressing facts in the latter realm cannot vary without variance of the truth values of sentences expressing facts of the former kind" (1975, p. 558). They formulate this basic idea with what they call their "Principles of Physical Determination" (1975, Sect. 2.1), which capture the sort of necessitation that supervenience theses aim to express. So, they propose, "[t]he physicalism that appears plausible has two components: ontological physicalism—the Principle of Physical Exhaustion—and Physical Determinationism" (p. 561).

their global or local versions, are *necessitation* claims. To capture the idea that mental facts depend entirely on physical facts, supervenience claims tell us that the physical facts *determine/necessitate* the mental facts. But a crucial question arises. What brand of necessity is at issue here? Is it logical necessity? Nomic necessity? Or is it some other variety? To know whether a supervenience thesis (conjoined with PD) captures the content of physicalism, this question needs to be answered.

2.3 Varieties of Necessitation

The physical details, the details regarding which individuals have which physical properties, clearly do not *logically* necessitate the mental details. The laws of logic alone do not enable us to derive the latter from the former. Yet, one might wonder whether the laws of logic together with the *meanings* of our words allow us to derive claims expressed in mental vocabulary from those expressed in physical terminology. One might wonder whether

($N_{L/C}$) facts about the distribution of physical properties and relations *logically/conceptually* necessitate all the facts about the distribution of mental properties/relations.

This way of stating the necessitation is meant to be amenable to global and local theses alike; it may be construed as a claim about what the totality of the world's physical details necessitate or as a claim about what the physical condition of each individual entails. Yet, $N_{L/C}$ seems implausible at least on standard interpretations of 'physical'. Suppose we restrict the term 'physical' to properties described by the science of physics itself. Then the obvious difference in meaning between mental vocabulary and the vocabulary of physics seems to show that $N_{L/C}$ is false. Suppose we use the word 'physical' loosely enough to include properties of the natural sciences generally, and brain properties in particular. Even if mental properties were identical with brain properties, $N_{L/C}$ would still be highly dubious. Since mental talk is not synonymous with neuroscientific talk, mind-brain identity theorists rightly held their view, not as an analytic truth, but as a significant empirical hypothesis.[12]

Not only does $N_{L/C}$ seem to be false, it also appears to be more than what physicalism requires. It might be that mental phenomena depend entirely on physical phenomena, as physicalists regarding mentality believe, without there being a logical or conceptual tie between the two. Predicates that differ greatly in meaning and logical form might denote the very same property. So even if mental properties were identical with physical properties, it would not follow that facts about the latter logically or conceptually guarantee facts about the former.

---

[12] See, for example, Smart (1959) and Place (1956).

That mental differences require physical differences would seem to have much to do with the *laws of nature* that actually obtain. If mental properties are not physical, then which mental properties are exemplified would seem to depend not only on the physical properties exemplified but also on the causal laws that actually obtain, especially the psychophysical laws (the laws regarding which mental properties accompany which physical properties). So perhaps we should say that

(N_N) facts about the distribution of physical properties/relations *nomically* necessitate all the facts about the distribution of mental properties/relations.

Yet, while it is reasonable to think that the supervenience of the mental on the physical is a matter of the laws of nature that obtain, it is widely recognized that N_N is too weak to do all the work a physicalist would want from a supervenience thesis. The physicalist who appeals to supervenience is trying to acknowledge that mentality depends entirely on the way the world is physically without being committed to the view that mental properties are identical with physical properties. Suppose, then, that mental properties are not physical properties. Then the psychophysical laws are not purely physical laws; they are laws connecting physical properties with non-physical properties. So it is consistent with N_N that there is a possible world that is physically indistinguishable from the actual world, but with a very different distribution of mental properties, and even without any mentality at all. If a physical duplicate of the actual world with radically different mentality were possible, then clearly the mental facts that actually obtain would be a function of more than just the physical facts, contrary to physicalism.[13] So, against the idea that N_N captures the content of physicalism, there is Crane's point that "if fixing the mental facts requires psychophysical laws, then fixing the physical facts *alone* is not sufficient to fix the mental facts" (1991, p. 237).

To capture physicalist intuitions, it seems we need a supervenience thesis that entails:

(N_P) facts about the distribution of physical properties/relations *physically* necessitate the facts about the distribution of mental properties/relations,

where *p* physically necessitates *q* just in case '*p* and not *q*' is not true in any possible world with the same physical laws as those that actually obtain.[14] If the physical laws did allow worlds with the same distribution of physical properties but

[13] A radical mental difference in physical duplicates of the actual world is a consequence of Chalmers' naturalistic, according to which there are fundamental phenomenal laws that obtain alongside the fundamental physical laws. This view allows worlds that physically duplicate the actual world with all of the actual physical laws, but whose psychophysical laws differ in such a way that there is a radically different distribution of phenomenal properties, and perhaps without any instances of phenomenal features (a zombie world).

[14] And to allow the possibility of mental *extras*, we might restrict the claim to *positive* mental facts, as Chalmers (1996, p. 40) mentions, which allows a physical duplicate of the actual world that does not duplicate the actual world's negative facts (such as the fact that there are no immaterial items).

a different distribution of mental properties, then there would be a clear sense in which the mental facts are at least partly due to something other than physical facts. So it seems that our physicalist intuitions demand just the type of necessitation that N_P describes.

One might wonder about the prospect of combining N_P with a claim about logical/conception necessitation. Imagine the complete set, *S*, of facts about the world expressed in the vocabulary of an ideal physics, and suppose this set includes all of the physical laws that obtain. According to Kirk, physicalism requires that the relation between *S* and the set of mental facts is one of *strict implication*, where a "statement *A* strictly implies a statement *B* just in case '*A* and not-*B*' involves inconsistency of a broadly logical or conceptual kind" (2006, p. 525). If Kirk were correct, then the physical necessitation described in N_P would itself hold as a matter of logical/conceptual necessity. There is a big debate in the literature about what does and does not follow *a priori* from the physical facts.[15] I wish to avoid this issue here and merely note that one certainly can earn the label 'physicalist' without believing that N_P is true as a matter of logical/conceptual necessity. It would be a mistake, then, to insist on this robust brand of necessitation in a definition of physicalism.

So perhaps we should rest modally content with mention of only physical necessity in N_P. Or if we wish to speak of *metaphysical* necessity, then we can say that in all metaphysically possible worlds with the same distribution of physical properties/relations *and all the same physical laws*, the same mental facts obtain. So, if one prefers, N_P may be replaced with:

(N_p*) facts about the distribution of physical properties/relations together with the physical laws metaphysically necessitate the facts about the distribution of mental properties/relations.

Unlike N_N, N_P and N_P* capture the physicalist idea that what necessitates the mental facts are facts that are purely physical (the distribution of physical properties/relations and the purely physical laws). Suppose we also endorse the Principle of Physical Decomposition (PD) to ensure that mentality is not only necessitated physically, but also instantiated physically. Then, it might be thought, we have all that our physicalist intuitions demand.

Unfortunately, the road to non-reductive physicalism is not quite so smooth. In the next section, I present a major worry for the non-reductive physicalist that has not been adequately recognized or addressed in the philosophical literature.

---

[15] Jackson (1994, 1998) and Chalmers (1996) contend that much more follows a priori from the physical facts than we might be inclined to think. Although, for Chalmers and Jackson, these do not include facts about the qualitative character of conscious experience, which they think is a reason to reject physicalism regarding the qualitative character of consciousness. (Although Jackson, e.g., 2003, 2012, subsequently embraces physicalism).

2.4 The Major Concern for the Non-Reductive Physicalist

Kim points out that a supervenience thesis "itself says nothing about the *nature* of the dependence" of the mental on the physical; "it tells us neither what kind of dependency it is, nor how the dependency grounds or explains the property covariation" (1993, pp. 165–166). Thus, a supervenience thesis all by itself leaves us wondering:

> Is it a matter of causal dependence? Is it in some way analogous to mereological supervenience? Is it after all a matter of meaning dependence, as logical behaviorists and some functionalists claim? Perhaps, a matter of divine intervention or plan as Malebranche and Leibniz thought? Or a brute and in principle unexplainable relationship which we must accept "with natural piety," as some emergentists used to insist? (1993, p. 167)

The worry that a mere supervenience claim leaves the nature of the mental-physical covariance unexplained is a common concern with defining physicalism in terms of supervenience.[16] The reason unexplained mental-physical covariance is worrisome is that, as physicalists, we would hope not only that there's an explanation of the covariance, but also that there's an explanation that preserves our physicalist commitments. As physicalists, we would hope for what Horgan calls "superdupervenience"—"ontological supervenience that is robustly explainable in a materialistically acceptable way" (1993a, p. 577).[17] And one good reason to desire a materialistically acceptable explanation of the dependence is that without such an explanation we could not be sure that the mental really does depend on *nothing other than* the physical.[18] So for those non-reductive physicalists who aim to define physicalism in terms of supervenience (instead of identity), the following worry is a common one:

(*I*) with a mere supervenience thesis, the covariance of mental properties with physical properties is left unexplained.

While ***I*** has been a major concern with supervenience-based accounts of physicalism, it seems to me that there is no fatal worry here simply because ***I*** is not entirely true. It should be clear from the presentation in Sect. 2.3 that deciding which brand of necessity should figure in a supervenience claim can take us a long way toward explaining why mental properties covary with physical properties in the way that physicalism requires. Suppose it were the case that mental properties supervene on physical properties as $N_{L/C}$ describes. Then there would be an easy explanation of why physical sameness guarantees mental sameness; the laws of logic and the meanings of our mental and physical terms would be wholly

---

[16] Others, besides Kim, who have expressed this concern include Gardner (2005, pp. 201–205), Heil (1998), Horgan (1993a, Sect. 8 and 1993b, Sect. 5), and Moreland (1998, pp. 49–53). See also Kim's 1989, Sect. 4.

[17] Although, Horgan (1993a, Sect. 8 and 1993b, Sect. 5) gives reason to doubt that there is such an explanation.

[18] Without an explanation we could not be sure simply because an explanation is meant to help us understand why some fact obtains (in this case, why the mental is fixed by the physical).

responsible. So the demand for superdupervenience would be satisfied with $N_{L/C}$. Given the implausibility of $N_{L/C}$ it is fortunate that N_P (or N_p*) also explains the covariance of mental properties with physical properties in a physicalistically acceptable way. If N_P is true, then the explanation of the covariance is that for any mental property M and any physical property P on which M supervenes, the physical laws that actually obtain are what make it the case that M is instantiated whenever P is. One might insist on further explanation, for the question does remain why it should be that the physical laws necessitate the psychophysical law, P → M, given that M is not identical with P; as Horgan mentions, "there remains the need to explain why these bridge laws *themselves* are true in all physically possible worlds" (1993a, p. 578). Yet, to be told that the physical facts fix the mental facts due to purely physical laws is certainly some explanation of the mental-physical covariance; it does at least answer the questions Kim poses in the quote above. Granted, "it is hard to see," as Horgan mentions, "how one could possibly explain inter-level supervenience facts without allowing oneself *some* sort of information about higher-level properties, concepts, or terminology" (1993b, p. 314). Yet, the explanation that N_P and N_P* provide seems explanation enough for the truth of physicalism as an *ontological* doctrine; the mental facts might be solely a function of the physical facts (as N_P and N_P* entail) even without a full explanation, or even a satisfactory explanation, of the covariance.[19]

There is, however, a related but distinct concern for the non-reductive physicalist, one which seems to me more troublesome. According to physicalism, the mental facts are solely a function of the physical facts, and as explained in 2.3 it seems that physicalism requires the mental facts to be fixed by the physical facts in the manner expressed by N_P (or N_p*), i.e., with the distribution of physical properties/relations together with the physical laws making it impossible that those mental facts fail to obtain.[20] Now if mental properties are identical with physical properties, then it is easy to see why there should be this sort of necessitation; if mental properties are physical, then mental facts are physical facts, and every fact trivially necessitates

[19] Our understanding of any of the necessitation claims described in 2.3 requires a grasp of mental concepts. This does not prevent the grounding of the mental by the purely physical as, for example, N_P and N_P* claim. Also, despite the fact that a grasp of mental concepts is required for comprehension of N_P and N_p*, it might be that these claims these claims nevertheless provide a sufficiently full explanation of the dependence of the mental on the physical. Gardner writes: "Since no P–N law can be inconsistent with or violate any purely physical law, there does not seem to be anything in virtue of which the set of actual physical laws can make it impossible for a given P–N relation to fail to obtain," and he concludes that "if there are P–N laws that are necessitated by the physical laws of the actual world ..., then this necessitation is wholly without explanation and the P–N laws are thus unacceptably brute" (2005, p. 202). However, there is another conclusion that might be drawn from the fact that P–N laws must be consistent with the physical laws. One might conclude that the P–N laws cannot violate the physical laws precisely because they are made true in virtue of nothing other than the physical laws, in which case, the P–N laws are not unacceptably brute.

[20] Physicalism at least requires that the physical facts fix the *positive* mental facts.

itself. But suppose that mental properties are not physical. What reason would there be in that case to expect the physical facts to fix the mental facts?

It's doubtful that there would be any reason to expect the physical to necessitate the mental. To see why it's doubtful, imagine a case of *absent mentality*. This is a possible scenario in which,

> (Absent) there is the same distribution of physical properties/relations as what actually obtains, and all the same physical laws, but some of the actual mental episodes are absent.

According to physicalism, the mental facts (at least the *positive* ones) are necessitated by the physical facts, including the physical laws; for if the mental facts were not necessitated by the physical facts, mentality would be a function of something more than the physical facts.[21] So, it seems, the actual truth of physicalism does not allow even the possibility of Absent cases.

However, if we thought that mental properties are not physical, then it's not clear why we should consider Absent cases impossible. Suppose P is a physical property that actually underlies M. If mental properties are not physical, then it seems there is a possible world, physically indistinguishable from the actual world, in which P is instantiated without M. One might suspect that the physical laws would prevent the absence of M in the presence of P. But on the assumption that mental properties are non-physical, the physical laws do not range over mental properties. Perhaps M is a consequence of P and the other physical details (including the physical laws), but being non-physical, M would not itself be part of the physical details. So why, then, should we deny that there is another possible world with all those physical details but without the instantiation of M?

Consider a world that contains what Hawthorne (2002) calls a "blocker." This is a world in which some non-physical entity (e.g., a god or spirit of some sort) blocks an instance of some physical property from having its typical mental consequence. Without changing any physical operations, the "blocker" prevents the physical event from yielding a mental episode of the sort normally produced by physical events of that type. Leuenberger (2008, p. 147) also has us imagine a mental episode being blocked. Consider a world where on day eight, after setting all the physical facts and having a day of rest, God infused someone's foot with "algoplasm," which has the effect of making phenomenal properties disappear. If mental properties were physical properties, then keeping the physical details the same

---

[21] Again, I speak of positive mental facts to allow that physicalism is compatible with the possibility of mental extras.

[22] Leuenberger (2008) claims that physicalism is compatible with the possibility of blockers, and proposes what he calls "*ceteris absentibus* physicalism," according to which, "[t]he actual physical facts are *ceteris absentibus* sufficient for all actual facts" (p. 148). On this account, if a world physically duplicates the actual world, then *other things being absent* (which includes no blockers) it mentally duplicates the actual world. However, since the other things being absent includes properties that are not purely physical, it seems that *ceteris absentibus* physicalism does not fully capture the physicalist belief that the mental is solely a function of the physical.

could not result in these mental absences. However, if mental properties are not physical, then it is not at all clear why these blocker cases should be impossible.[22] The concern is not that the necessitation lacks an explanation. The question is why we should think the necessitation even obtains (explicable or not) if we deny that mental properties are physical.

The notion of a blocker provides a reason to suspect that Absent cases are possible, i.e., reason to suspect they are possible on the assumption that mental properties are not physical. But there are Absent cases that do not involve blockers; consider those zombie scenarios where consciousness is absent in physical duplicates of us but without any blockers involved. Now, if sameness of physical details is consistent with mental absences in worlds with immaterial blockers, then there seems no reason to deny that sameness of physical detail is also consistent with mental absences and no blockers. It may be that Absent cases require the violation of actual psychophysical laws, and if they do, then these cases do not obtain in any *nomically* possible worlds. But if mental properties are not physical, then it's not clear why we should deny that they obtain in some *physically* possible worlds, even without intervening blockers. Suppose that instances of some mental property M accompany instances of physical property P in all nomically possible worlds. Unless M itself is a physical property, there seems no reason to deny the possibility of a creator, intelligent or not, that produced a physical duplicate of the actual world with all of the actual physical laws (though not all the same psychophysical laws), in which instances of P are not, or at least not always, accompanied by instances of M. In this case, the creator is not blocking the mental episodes at the time, but rather has designed a world that physically duplicates the actual world but where instances of M do not always accompany instances of P as they do at the actual world.

The worry that is being expressed here is not merely that Absent scenarios are logically or conceptually possible. It seems clear that the physical facts do not fix the mental facts solely as a matter of logic and definition; moreover, the idea that Absent cases are logically/conceptually possible is perfectly compatible with physicalism since mental facts can obtain solely in virtue of the physical facts, as physicalism demands, without doing so analytically. The real worry for the non-reductive physicalist is that if mental properties are not physical, then Absent cases seem to be *metaphysically* possible, i.e., genuine possibilities, and this does conflict with the physicalist view that the mental obtains solely in virtue of the physical. It seems, then, that the non-reductive physicalist faces the following worry:

(***II***) mental properties not being identical with physical properties prevents the physical facts from necessitating the mental facts in the way that physicalism requires.

Note that this concern for the non-reductive physicalist is different from ***I***. There are supervenience theses, which if true would explain, at least to some degree, why mental properties covary with physical properties. N_P is one such supervenience thesis, and it seems just the sort of supervenience a physicalist would demand. Yet, even with explanatory concern ***I*** aside, the worry remains that there's reason to doubt the physical necessitation (explicable or not) expressed by N_P if mental properties are not physical.

The preceding points are not meant to support the possibility of Absent cases. For all that has been said here it might be that mental properties are in fact physical, and if so, then given the necessity of identity, Absent cases are not possible. It might also be that Absent cases, for some non-obvious reason, are impossible even with mental properties not being physical. The point here is simply that it is quite unclear why we should think they are impossible if we deny that mental properties are physical. This is a worry for the non-reductive physicalist since the possibility of Absent cases seems to be incompatible with the actual truth of physicalism. For if we allow that the physical details can be exactly the same without all the same (positive) mental facts, then it seems we forfeit the idea that the mental facts obtain solely in virtue of the physical facts.

The worry for the non-reductive physicalist may be expressed as follows. (i) If physicalism is true, then Absent cases are impossible. (ii) If mental properties are not identical with physical properties, then Absent cases are not impossible. So (iii) if mental properties are not identical with physical properties, then physicalism is not true. Assuming that (i) is true, as it seems to be,[23] the proponent of non-reductive physicalism owes some good reason for thinking that (ii) is false, i.e., a good reason for thinking that ***II*** is false. Of course, showing that ***II*** is false does not require showing that physicalism is in fact true, but only that non-reductionism does not itself prevent physicalism from being true.[24]

Since the rise of non-reductive physicalism, much talk has been made of mental phenomena being *realized* by physical phenomena. In the next chapter, we will consider whether the appeal to realization helps to show why ***II*** should be rejected.

References

Bennett, K. 2004. Global supervenience and dependence. *Philosophy and Phenomenological Research* 68: 501–529.

Chalmers, D. 1996. *The conscious mind.* New York: Oxford University Press.

Crane, T. 1991. All God has to do. *Analysis* 51: 235–244.

Davidson, D. 1970. Mental events. In *Experience and theory*, ed. L. Foster, and J.W. Swanson, 79–101. Amherst: University of Massachusetts Press.

Francescotti, R. 2000. Ontological physicalism and property pluralism: Why they are incompatible. *Pacific Philosophical Quarterly* 81: 349–362.

---

[23] But recall Leuenberger's "*ceteris absentibus* physicalism" mentioned in the previous footnote.

[24] In an earlier work, I (2000) argue that physicalism is incompatible with the denial of mental-physical property identities. I realize that the reasoning there was hasty; yet it nonetheless seems quite true that to justify an allegiance to physicalism the non-reductionist owes some explanation of how one can be entitled to believe that physicalism is true (and therefore that Absent cases are impossible) if one believes that mental properties are not physical.

Francescotti, R. 2014. The problem of extras and the contingency of physicalism. *Philosophical Explorations* 17: 241-254
Gardner, T. 2005. Supervenience physicalism: Meeting the demands of determination and explanation. *Philosophical Papers* 34: 189–208.
Haugeland, J. 1984. Ontological supervenience. *The Southern Journal of Philosophy, Supplement* 22: 1–12.
Hawthorne, J. 2002. Blocking definitions of materialism. *Philosophical Studies* 110: 103–113.
Heil, J. 1998. Supervenience deconstructed. *European Journal of Philosophy* 6: 146–155.
Hellman, G., and F. Thompson. 1975. Physicalism: Ontology, determination, and reduction. *The Journal of Philosophy* 72: 551–564.
Horgan, T. 1982. Supervenience and microphysics. *Pacific Philosophical Quarterly* 63: 29–43.
Horgan, T. 1993a. From supervenience to superdupervenience: Meeting the demands of a material world. *Mind* 102: 555–586.
Horgan, T. 1993b. Nonreductive materialism and the explanatory autonomy of psychology. In *Naturalism: A critical appraisal*, ed. S.J. Wagner, and R. Warner, 295–320. South Bend, IN: University of Notre Dame Press.
Jackson, F. 1994. Armchair metaphysics. In *Philosophy in mind*, ed. M. Michael, and J. O'Leary-Hawthorne, 23–42. Dordrecht: Kluwer.
Jackson, F. 1998. *From metaphysics to ethics: A defence of conceptual analysis*. Oxford: Clarendon Press.
Jackson, F. 2003. Mind and illusion. In *Minds and persons*, ed. A. O'Hear, 251–271. Cambridge: Cambridge University Press.
Jackson, F. 2012. In defence of the identity theory mark I. In *New perspectives on type identity: The mental and the physical*, ed. S. Gozzano, and C. Hill, 150–166. Cambridge: Cambridge University Press.
Kim, J. 1984. Concepts of supervenience. *Philosophy and Phenomenological Research* 45: 153–176.
Kim, J. 1987. "Strong" and "global' supervenience revisited. *Philosophy and Phenomenological Research* 48: 315–326.
Kim, J. 1988. Supervenience for multiple domains. *Philosophical Topics* 16: 129–150.
Kim, J. 1989. The myth of nonreductive materialism. *Proceedings and Addresses of the American Philosophical Association* 63: 31–47.
Kim, J. 1990. Supervenience as a philosophical concept. *Metaphilosophy* 21: 1–27.
Kim, J. 1993. *Supervenience and mind: Selected philosophical essays*. Cambridge: Cambridge University Press.
Kirk, R. 2006. Physicalism and strict implication. *Synthese* 151: 523–536.
Leuenberger, S. 2008. *Ceteris absentibus* physicalism. In *Oxford studies in metaphysics*, vol. 4, ed. D. Zimmerman, 145–170. Oxford: Oxford University Press.
Leuenberger, S. 2009. What is global supervenience? *Synthese* 170: 115–129.
Lewis, D. 1983. New work for a theory of universals. *Australasian Journal of Philosophy* 61: 343–377.
Lewis, D. 1986. *On the plurality of worlds*. Oxford: Basil Blackwell.
McLaughlin, B. 1995. Varieties of supervenience. In *Supervenience: New essays*, ed. E. Savellos, and U. Yalcin, 16–59. Cambridge: Cambridge University Press.
McLaughlin, B. 1997. Supervenience, vagueness, and determination. *Philosophical Perspectives* 11: 209–230.
Moreland, J.P. 1998. Should a naturalist be a supervenient physicalist? *Metaphilosophy* 29: 35–57.
Moyer, M. 2008. Weak and global supervenience are strong. *Philosophical Studies* 138: 125–150.
Paull, C., and T. Sider. 1992. In defense of global supervenience. *Philosophy and Phenomenological Research* 52: 833-854.
Petrie, B. 1987. Global supervenience and reduction. *Philosophy and Phenomenological Research* 48: 119–130.
Pettit, P. 1993. A definition of physicalism. *Analysis* 53: 213–223.

Place, U.T. 1956. Is consciousness a brain process? *British Journal of Psychology* 47: 44–50.
Shagrir, O. 2009. Strong global supervenience is valuable. *Erkenntnis* 71: 417–423.
Shagrir, O. 2013. Concepts of supervenience revisited. *Erkenntnis* 78: 469–485.
Smart, J.J.C. 1959. Sensations and brain processes. *The Philosophical Review* 68: 141–156.
Stalnaker, R. 1996. Varieties of supervenience. *Philosophical Perspectives* 10: 221–241.
Yoshimi, J. 2007. Supervenience, determination, and dependence. *Pacific Philosophical Quarterly* 88: 114–133.

Chapter 3
Realization and Constitution

Physicalism regarding the mind seems to entail that mental properties supervene on physical properties. In particular, physicalism seems to require that the distribution of mental properties is necessitated by the distribution of physical properties together with the physical laws that obtain. However, as was illustrated in Sect. 2.4 with the possibility of absent mentality cases, it is not clear what would justify the belief that mental properties supervene on physical properties in this way if one denies that the former are identical with the latter. The worry, we saw, is what I labeled "***II***"—the worry that mental properties not being identical with physical properties prevents the physical facts from necessitating the mental facts in the way that physicalism requires. To be convinced that non-reductive physicalism is a viable position, we need to be shown why ***II*** is false, where showing that ***II*** is false does not require showing that physicalism is in fact true, but only that non-reductionism does not itself prevent physicalism from being true.

Since the rise of non-reductive physicalism much talk has been made of mental phenomena being *realized* by physical phenomena. Perhaps the idea that mentality is realized physically reveals why ***II*** is false, thereby showing how one can remain faithful to physicalism despite being a non-reductionist. In this chapter, we will consider whether this is the case.

In Sects. 3.1 and 3.2 the basic idea of realization is reviewed, with a discussion in Sect. 3.3 of Melnyk's rigorous formulation of the notion. In Sect. 3.4 we consider Shoemaker's subset account of realization, and in Sect. 3.5 the type of physical realization Pereboom calls "material constitution" is discussed. It is argued that none of these accounts of realization shows why ***II*** is false. And if this conclusion is correct, then none of the accounts of realization discussed here adequately reveals how one can justifiably accept physicalism while denying that mental properties are physical.[1]

---

[1] The ideas in Sects. 3.1–3.3 have been published elsewhere, in Francescotti (2010). They are presented here in revised form.

R. Francescotti, *Physicalism and the Mind*,
SpringerBriefs in Philosophy, DOI 10.1007/978-94-017-9451-0_3

3.1 Functional Properties and Realization: The Basic Idea

A functional property is a property an item exemplifies in virtue of playing a certain functional role. That is,

> F is a functional property = there is a functional role, R, such that for any x, x's exemplifying F consists in x's playing role R.

Being a *soda-machine*, for example, is a functional property since there is a certain function—the function of dispensing beverages when the right coins are inserted and tabs pressed—and being a soda machine consists in being a machine that performs (or was designed to perform) this beverage-dispensing function.[2] Being a heart is also a functional property since an object exemplifies this property if and only if it plays (or evolved to play) the right sort of blood-pumping role. It is also commonplace, thanks to the popularity of functionalism in the philosophy of mind, to think of mental properties as functional. To be in pain, we are told, is to have some inner event that plays the right causal role with respect to sensory input (e.g., cuts, burns. and pressure), behavioral output (grimacing, wincing, groaning), and other mental states (beliefs about the source of the pain and the desire to avoid that source).

With the notion of a functional property comes the notion of a *realization* of that property. In general, for any functional property, F, what plays the functional role definitive of F is called a "realization" of F. The large object in the hallway is a realization of the property of being a soda-machine in virtue of playing the role of dispensing soda, and the pumping thing in your chest realizes a heart in virtue of serving to pump blood. In general,

(a) property F is *realized* by item x at time t = there is a functional role, R, such that exemplifying F consists in playing R, and x plays R at t.

The word 'item' includes any concrete particular. It might be an *object*, as when a physical structure realizes the property of being a table, heart, or soda-machine. Or it might be an *event* or *process*, as when instances of different types of movement, e.g., walking and running, realize the same behavioral property of locomoting or when events of different neural types realize the same mental property.

If some property, F, is a functional property, then whether an item exemplifies F depends entirely on whether the object plays (or was designed to play) the right functional role. The internal composition matters to whether an object is F only to the extent that the inner composition enables the object to perform the right function. It is no surprise, then, that functional properties are *multiply realizable*. Two objects might be comprised of different types of metal but function equally well as a soda-machine, or one object comprised of wood and another made of plastic might both perform the function of a table.

---

[2] Recall Block's (1978) classic functional description of a soda-machine.

The definitions above treat functional properties as *first-order* properties, but we might wish to view them as *second-order* instead. Rather than saying that having functional property F consists in playing role R, we might prefer to say that having F consists in *having a property that plays R*. In that case, F is viewed as the property of having a property that plays R, which is a second-order property. As Block (1990, pp. 155, 163) illustrates, dormitivity may be viewed simply as the first-order property of causing sleep or it may be considered the second-order property of having a property that causes sleep.

In the philosophical literature mental properties are often described as second-order. Van Gulick states that "having a given mental property M is a matter of having some set of other properties P_1 ,..., P_n that satisfy the functional relation R_M," where 'R_M' specifies the functional role definitive of M (1992, p. 164). Papineau proposes that "[i]n order for a mental or other special type M to be realized by an instance of some physical type P, M needs to be a *second-order property*, the property of having some property which satisfies certain requirements R" (1993, p. 25). Block describes functional properties as "properties that consist in the having of some properties or other (say non-functional properties) that have certain causal relations to one another and to inputs and outputs" (1990, p. 155). If we wish to follow their lead and construe functional properties as second-order, we might choose the definition:

> F is a functional property = there is a functional role, R, such that for any x, x's exemplifying F consists in x's *having a property that plays R*.

If we think of realized properties as second-order, then it is natural to view the realizers as properties as well. The realizer of second-order property F will be one of the first-order properties that plays R. Thinking of functional properties as second-order and their realizers as first-order properties that play the right functional role yields the following account of realization:

(b) property F is realized by property G for individual x at time t = there is a functional role, R, such that exemplifying F consists in having a property that plays R, and G plays R for x at t.

Definitions (a) and (b) mention functional roles rather than *causal* roles to allow that some functional properties are non-causal. We might wish to view mental properties as *computational* rather than causal, which is why Melnyk (1996, p. 391) mentions that in addition to "causal-functional" properties, there are *computational-functional properties*. What allows for realization in this case is not the causal role played by the object's states, but the *isomorphism* between the object's states and states of a program. To illustrate a third type of functional property, Melnyk notes the property of being a heart. Being a heart consists in having a certain biological function—the biological function of pumping blood. But suppose the heart has been damaged and doesn't adequately perform that function. We would still wish to call it a heart because we know that there is a function that all hearts, including this damaged one, have *evolved* to perform. Understanding the function an organ evolved to perform obviously requires understanding its *ancestry*, and Melnyk points out, "if this notion of ancestry cannot itself be spelled out causally, then we will have to acknowledge a third type of functional property, which may be called functional-functional" (1996, p. 392). One might wonder whether these "functional-functional" properties can be reduced to causal properties, whether each causal property can be construed as a computational property, and whether there are other major brands of functional property in addition to these three. I shall not address these issues here. Suffice it to say that there may be reason for viewing the role definitive of some functional properties as something other than causal, which is why definitions (a) and (b) do not mention causal roles.

Melnyk believes that mental properties are not identical with physical properties, yet he still considers himself a physicalist. He proposes that "a doctrine of retentive physicalism can be formulated using only the notion of realisation; neither claims of supervenience nor claims of token identity nor disjunctive definitions … play any part whatever" (1996, p. 390). On Melnyk's understanding of physicalism, "[e]very property exemplified in the actual world is either (i) a physical property or (ii) a functional property of one of the allowed types and physically realised *simpliciter*, either directly or indirectly" (p. 395). If physicalism is to be defined in terms of realization, as Melnyk proposes, then the doctrine ends up being perfectly compatible with *property dualism*.[3] According to the basic notion of realization, expressed as either (a) or (b), what realizes a mental property M on one occasion might be physically very different from what realizes the same mental property on some other occasion. Also, the differences in realization might be such that there is no one property of physics, or any other natural science, that is necessary and sufficient for the presence of M,[4] in which case, M is not identical with a property of physics or any other natural science.

It is clear, then, that the mental can be realized by the physical, in either sense (a) or (b), even though mental properties are not themselves physical. However, as shown in the next section, the appeal to realization, of type (a) or (b), does not by itself capture the content of physicalism. In particular, an account of physicalism in terms of realization, of type (a) or (b), fails to reveal why ***II*** is false. In Sect. 3.3, I argue that the limitations of "Realization Physicalism" described in Sect. 3.2 also apply to Melnyk's more sophisticated (2006) analysis of realization.

---

[3] Applied to the mental and the physical, property dualism is the view that mental properties are not identical with physical properties, which in itself allows the truth of physicalism—assuming, that is, that non-reductive physicalism is a coherent position.

[4] This would be an instance of *robust* multiple realizability, as defined in Chap. 1 (and robust_p with the absence of physically necessary and sufficient conditions, and robust_m with the absence of metaphysically necessary and sufficient conditions).

3.2 The Mental Realizing the Physical

According to Melnyk's Realization Physicalism, "[e]very causal or contingent token of any type … is either (1) a token of a physical type or (2) a physically realized token of a functional type" (2006, p. 131). If this view is to have any chance of being true, Melnyk had better be using the word 'functional' quite broadly, to include all non-physical types; and he does admit that his "account uses 'functional type' very liberally indeed, to refer to any type whose tokening just is the tokening of some or other type that meets a specific associated condition, where the condition could be of *any* kind" (2006, p. 129).

Now, in Melnyk's broad sense of 'functional', functional properties are abundant in the realm of physics as well as the other natural sciences. For many if not most physical types, there is an associated functional role that must be played in order to have a token of that type. An item is an *electron* just in case it plays the role of being a *subatomic particle with negative electric charge*. And associated with the physical kind, *boson*, is the functional role, *particle with integer spin obeying Bose-Einstein statistics*, and something is a *boson* if and only if this condition is met. Since these physical properties qualify as functional types, physical properties can be realized in sense (a) and (b). This alone is no threat to Realization Physicalism since the claim that all concrete phenomena depend entirely on physical phenomena does not prevent physical properties themselves from being realized physically. However, the question does arise: if some physical properties are functional in character, and therefore realizable, can they also be realized by items that are non-physical? For instance, can physical properties be realized by mental items?

It seems the answer is "yes." Consider the very general physical property of *having a certain density*. This qualifies as a functional property, at least in the loose sense of 'functional property' Melynk describes, according to which, a functional property is one whose instantiation consists in the satisfaction of a condition of any kind. There is a functional role, R, such that an item has density just in case the item plays R, where R is the property of *having some ratio of mass to volume*. Although Cartesian egos, and immaterial souls generally, seem to be logically and even metaphysically possible, it is not implausible to think, as physicalism demands, that as a matter of fact all sentient organisms are bodies in physical space and therefore have a certain density. So assuming that substance dualism is false, each sentient organism plays the role of having a certain ratio of mass to volume, which is definitive of having density. Thus, each sentient organism realizes the property of having density according to (a). Having density is a genuine property and a genuine property of physics, and being sentient certainly qualifies as a mental property. So in this case a physical property is realized by mental items (sentient organisms) in manner (a).

In this case the realizer is a concrete individual item—the sentient organism. If we think of realization in terms of (b), then we will view realizers as properties,

properties playing the roles definitive of higher-order functional properties. Yet, the property of being sentient can qualify as a realizer according to (b) just as much as the sentient organism can qualify as a realizer given (a). Having density may be construed as the second-order property of *having a property that plays role R*, where R is *instantiated in an item with a certain ratio of mass to volume*. The mental property of *being a sentient organism* plays role R, and in virtue of doing so it realizes the property of having density, according to (b). Thus, whether we endorse (a) or (b)—whether we think of functional properties as first- or second-order, and whether we think of realizers as concrete particulars or as properties—we get the conclusion that the mental can realize the physical.

This instance of mental realizing physical is not an isolated case. In addition to having a certain density, there are a number of very general physical properties that are realized by mental items. These physical types include: *having weight*, *having potential energy*, and *containing fermions*. Each of these is a general feature of physics. And each may be considered a functional type since having these features consists in a certain role being played. For the three physical types just mentioned, there is the role: *is subject to gravitational force*, *has stored energy*, and *contains elementary particles with half-integer spin*, respectively. Assuming that substance dualism is false, each thinking or feeling organism plays these roles and by doing so has weight and potential energy and contains fermions. So in each of these cases, mental items (thinkers and feelers) realize a clearly physical property according to (a). And regarding (b), let the realizer be the property of *being sentient*, and let the realized property be the higher-order property of *having a property that plays role R*, where R is *being instantiated in an item that* _____, with the ellipsis filled in by 'is subject to gravitational force', 'has stored energy', or 'contains elementary particles with half-integer spin'.

The physical types mentioned so far are extremely general in character. However, mental realizing physical in the sense of (a) and (b) also occurs in the case of highly specific physical properties. Take any type of color sensation, M, and any instance *e* of M. Assuming that physicalism is true, whenever M is instantiated, M is instantiated in some physical process or other. Now, suppose that P is the physical property that mental event *e* instantiates, and let R be the functional role of *being an instance of P*. Then mental event *e* realizes physical property P, according to (a), since it plays the role of being an instance of P (which is trivially what having P consists in). So here we have a highly specific physical property being realized by a mental item (color sensation *e*) according to (a). And suppose we let P* be the higher-order property of having a property that plays role R, where R is *being instantiated P-ly*. P* is clearly a physical property, given that P is. So in this case mental property M realizes a physical property, according to (b). Thus, whether we think of realization in terms of (a) or in terms of (b), we must acknowledge that highly specific physical properties can be realized mentally.

One might wonder whether in each of these cases, the role R that is described qualifies as a genuine *functional* role, for one might think that the role must be sufficiently causal or perhaps adequately purposive to deserve the label 'functional'.

However, as Melnyk recognizes, if Realization Physicalism is to have any chance of being true, the word 'functional' will need to be construed very liberally, since on his view, all non-physical types with contingent tokens qualify as functional types (functional types ultimately realized physically). And it is arguable that some non-physical types are not especially causal or purposive; some of them might even be considered *intrinsic* features of the items that bear them, and therefore not at all a matter of how the items relate to distinct things. One certainly can earn the title 'physicalist' while allowing that some mental (or otherwise non-physical) properties are intrinsic and therefore not functional in any robust sense. So to have even a chance of success, the word 'functional' in the formulation of Realization Physicalism had better, as Melnyk recognizes, be used quite liberally indeed.[5]

The fact that there are cases in which physical properties are realized mentally would seem to undermine the prospect of characterizing physicalism in terms of realization. Physicalists believe that all concrete phenomena occur solely in virtue of physical phenomena. But if a mental item can realize a physical property, then it seems that whatever makes it the case that mentality obtains in virtue of physical phenomena, it is not simply that the former is realized by the latter. So, it seems, the fact that mentality is realized physically is not enough to guarantee that the mental is grounded in the physical, and therefore not enough to capture our physicalist intuitions.

The fact that the physical realizes the mental is not itself what makes it the case that the latter obtains in virtue of the former. But, one might suggest, what makes it the case that the mental obtains in virtue of the physical is that the physical realizes the mental together with the fact that the *ultimate realizers* are physical and not mental. That is, for any mental property, there is a level of physical structure at which the mental property is realized physically and the physical realizer is not itself realized mentally. That's what makes it the case that the mental is dependent on the physical and not the other way around—or so one might contend.

But it is simply not true that there's a physical level that's not realized mentally. Suppose that P is a fundamental physical property (or highly fundamental if there is no most fundamental level). And suppose that M is some mental property that supervenes on P, and that an individual has P and thereby has an instance *e* of M. As before, let R be the functional role of *being an instance of P*. Then even though P is a fundamental physical property, mental event *e* realizes P according to (a), since *e* plays the role of being an instance of P. Also, as before, let P* be the higher-order property of having a property that plays role R, where R is *being instantiated P-ly*. Assuming that P is fundamental, P* is as fundamental as a higher-order property can be. And clearly P* is physical since P is. So in the case we are considering, mental property M realizes a fundamental physical property according to (b).

[5] The associated "condition could be of *any* kind, and needn't be the playing of a causal role" (Melnyk 2006, p. 129).

It seems, then, that characterizing physicalism in terms of realization does not fully capture the content of physicalism. And, in particular, the idea that mentality is realized physically offers no help in showing that the worrisome ***II*** is false. Since mental properties can realize physical properties, and even fundamental physical properties, the fact that mentality is realized physically gives us no reason to believe that mental facts are necessitated by physical facts in the way that physicalism requires. Quite the contrary, since mental phenomena can and do realize even basic physical phenomena, the idea that mentality is realized physically does nothing to prevent not only the possibility, but the actuality of the world being ultimately purely mental! So it's consistent with Realization Physicalism (the idea that all concrete phenomena are realized by physical phenomena) that there are possible worlds, physically indistinguishable from the actual world but in which everything is ultimately mental. Thus, Realization Physicalism does not entail that the physical fixes the mental in the way that physicalism requires.

The analysis of realization Melnyk (2006) offers is much more sophisticated than (a) and (b). For that reason one might wonder whether this account avoids the result that physical properties (at least fundamental physical properties) can have mental realizers.

3.3 Melnyk's Account of Realization

According to (a), realizers are concrete particulars, either the objects that exemplify the properties or the processes that instantiate those properties. According to (b), realizers are properties, the properties that play the requisite functional roles. But either way, the thing that is being realized is a property.

While not the typical way to proceed, one might also view the realized items as concrete particulars (realized by other concrete particulars). For instance, we might speak of a particular refrigerator being realized by a certain physical object, or a certain instance of pain being realized by some instance of a neural property. This is how Melnyk thinks of realization—as relating tokens (particular instances) to other tokens. He offers the following analysis:

Token *x realizes* token *y* (or: token *y* is *realized by* token *x*) iff

i. *y* is a token of some *functional* type F (i.e., some type whose tokening just is the tokening of some or other type that meets a certain condition, C);
ii. *x* is a token of some type that in fact meets C; and
iii. the token of F whose existence is necessitated (in the strongest sense) by the holding of clause (ii) is numerically identical with *y*. (2006, p. 129)

Let's label this analysis "R_M," to indicate realization Melnyk-style.

Note that clause (iii) does not require that the realizer token, x, is *identical* with the realized token, y. What (iii) requires to be identical with y is the token of F whose existence is necessitated by x's being a token of a type that meets condition C, which does not entail that x itself is identical with this token of F. So, for example, neural event x that realizes my current pain, y, must be of a type that plays the right functional role, thereby necessitating an instance of pain, but x need not be identical with this instance, y, of pain. Without the token-identity (between x and y) Melnyk's analysis allows us to believe that mental events are realized by physical events and also that organisms with mental features are realized by physical bodies, while remaining neutral on the difficult issue of whether *constitution* counts as identity—e.g., neutral on whether the mental event is identical with the constituent neural event, or whether the person with mental properties is identical with the constituent body.[6]

Melnyk claims that the tokening of functional type F is necessitated *in the strongest sense* by the tokening of a type that meets condition C. In many cases the connection between a functional type and the associated condition is *conceptual*. It is part of the very meaning of 'soda-machine' that a soda-machine is designed to dispense containers of soda given the right input, and a *tendon*, by definition, is fibrous tissue designed to connect muscle to bone; and if analytic functionalists are correct, then it is a conceptual truth that mental properties are functional properties. Presumably, this is why Melnyk (2003, p. 21) requires that a token of a type meeting condition C *logically* guarantees a token of F (where 'logically' is understood broadly enough to include *conceptual* ties). He later decides to require only that "the tokening of some or other type that meets condition C *metaphysically* necessitates a tokening of functional type F" (2006, p. 129; emphasis added). With this weaker constraint, Melnyk wants to allow that the condition essential to a functional type might not be discoverable *a priori*—i.e., that there are *synthetically necessary* truths about the functional character of various properties. So Melnyk's qualifier, "in the strongest sense," in definition R_M should be viewed as including metaphysical necessity.

There is reason *not* to think, as Melnyk does, that *both* relata of the realization relation are tokens. While Melnyk's account does not require that the realizer is identical with the token that is realized, it does not and should not prevent it either. Yet, if token x is identical with token y, then thanks to the indiscernibility of identicals, if x realizes y, then y realizes x, which seems unacceptable. This is one implausible consequence of the idea that the realizer and the realized are both tokens. Another, and more obvious, implausible consequence is mentioned by Polger and Shapiro (2008, p. 214). They point out that property-instances, tokens, are not the things realized, for if they were, then *multiple* realizability would be impossible since only properties are multiply instantiated. However, it is not necessary to worry here about complaints regarding the realized and the realizer both

---

[6] One might think that a mental event is not identical with the constituent neural event for the same reason one might believe that a statue is not identical with the constituent lump of clay—namely, the mental episode (statue) has *different persistence conditions* than the neural event (lump of clay). For example, an item can continue to be the same neural event (lump of clay) while ceasing to be a mental episode (statue). We will have more to say about constitution and identity in Sect. 3.5 when discussing Pereboom's constitution view.

being tokens, for the unwelcome consequence that mentality realizes physical phenomena arises even if R_M were rephrased to mention types being realized.

Recall the very general physical type of *having a certain density*. This is a functional type, in Melnyk's sense of 'functional', since there is a condition, C, that must be met (a role to be played) for there to be a token of this type. Condition C in this case is, *instantiated in an item with a certain ratio of mass to volume*, and having density consists in having a property (being of a type) that meets this condition. Now let G be the mental type that includes all and only *organisms that are sentient*. Although immaterial souls seem to be logically possible, it is not implausible to think, as physicalism demands, that as a matter of fact all sentient organisms are material objects in physical space, and therefore have a certain density. So assuming that substance dualism is false, for any sentient being, x, there is a physical token, y, where y is the physical body of x, and (i) body y is a token of type F (a bearer of density), and (ii) x is a token of mental type G (sentient organism) that meets condition C—the condition of being tokened in an item with a certain ratio of mass to volume. Also, (iii) the fact that x is of a type that meets C necessitates that there is a token of F (since there being a token of a type that meets the condition of being tokened in something with a certain ratio of mass to volume necessitates that there is something with a certain density), and this necessitated token of F is identical with token y. Now, having density is a genuine property and a genuine property of physics, and being sentient certainly qualifies as a mental property. So in this case, a mental token realizes a physical token on Melynk's account. (And it is easy to see that R_M gives the same result for each of the other general physical types mentioned in Sect. 3.2—*having weight*, *having potential energy*, and *containing fermions*).

Moreover, like (a) and (b), R_M allows that mental realizes physical in the case of highly specific physical properties as well. Suppose that G is a highly specific mental property—highly specific to ensure that the underlying physical feature is also quite specific. Suppose, for instance, that G is believing that 7^2 is 49, hoping that Gilliam brings exactly 2 cases of beer, or a peculiar tingling sensation quite unlike anything you've experienced before. Consider the purely physical description of some instance x of mental type G, and let F be the complex property denoted by this physical description. Then x is a token of a mental type that meets the condition of *being instantiated in manner F*. Now x's being a token of a type that meets the condition of being instantiated in manner F trivially necessitates that there is a token y of type F. Since F is the way in which mental property G is instantiated at the physical level, F counts as a genuine physical type. So here we have a highly specific physical process (token y of F) being realized mentally (by token x of G) according to R_M.

As mentioned in the previous section, the fact that there are cases in which physical phenomena are realized by mental phenomena would seem to undermine the prospect of characterizing physicalism in terms of realization. Physicalists believe that all concrete phenomena occur solely in virtue of physical phenomena. But if a mental item can realize a physical property, then it seems that whatever makes it the case that mentality obtains in virtue of physical phenomena, it is not simply that the former is realized by the latter. It appears, then, that mentality being realized physically is not enough to guarantee that the mental depends on the physical, and therefore not enough to capture our physicalist intuitions.

One might try to save Realization Physicalism by adding that there is a level of subvening physical structure with respect to which there are no non-physical realizers. But as with (a) and (b) this option is not available here either. Again, consider the purely physical description of instance x of mental type G, and let F be the complex property denoted by this physical description. Suppose that the physical description that denotes F is at some highly fundamental level. Still, x is a token of mental type G that meets the condition of *being instantiated in manner F*, and x's being a token of a type that meets this condition entails that there is a token y of type F. So here we have an instance of a fundamental physical property being realized mentally according to R_M.[7]

And, as mentioned at the end of Sect. 3.3, if mental properties can realize physical properties, and even fundamental physical properties, then the fact that mentality is realized physically gives us no reason to believe that mental facts are necessitated by physical facts in the way that physicalism requires. In fact, given that mental phenomena can and do realize even basic physical phenomena, the idea that mentality is realized physically does nothing to prevent not only the possibility, but also the actuality that the world is ultimately purely mental. So, it seems consistent with Realization Physicalism—understood in terms of (a), (b), or R_M—that there are possible worlds, physically indistinguishable from the actual world, but in which everything is ultimately mental. If such worlds were possible, then the physical details of the actual world (including the physical laws) would certainly fail to necessitate the way the world is mentally. So whether we construe realization in terms of (a), (b), or R_M, the idea that mentality is realized by physical phenomena provides no relief from worry ***II***. We are still left wondering how it could be that the physical facts necessitate the mental facts if mental properties are not physical. The claim that mental phenomena are realized by physical phenomena, in manner (a), (b), or R_M, does not reveal the answer to that question. In particular, the realization claim offers no help to understanding why the Absent cases described in Sect. 2.4 should not be metaphysically possible (contrary to physicalism) if mental properties are not physical.

---

[7] It's worth mentioning that the non-physical realizing the physical is not found only in the case of mentality. If mental phenomena can realize the stuff of physics, then it should be no surprise that chemical and biological phenomena can do so as well. Consider, for example, a token x of the biological type, *tendon*, and suppose that x is instantiated in manner P at the level of physics. In this case, x is of a biological type that meets the condition, C, of being instantiated in manner P, and meeting this condition necessitates that there is a token of P. So according to R_M, and also (a), biological tendon-token x realizes the stuff of physics. [And if we think of the property of being a tendon rather than a particular tendon as the realizer, and if we also focus on the second-order property of *being a property instantiated in manner P*, we can see that the biological can realize the physical according to (b) as well].

Adding Physical Necessitation It should be noted that after giving his analysis of realization (definition R_M), Melnyk adds an account of what it is for a functional property to be *physically* realized. He claims that a

> token *y* of a functional type, F, is *physically realized* iff (i) *y* is realized ... by a token of some physical type, T; and (ii) T meets the special associated condition for F solely as a logical consequence of (a) the distribution in the world of physical tokens and (b) the holding of physical laws (2006, pp. 130–131).

Adding condition (ii) does not prevent the mental from realizing the physical, but it does allow us to insist, as physicalism requires, that the realization of mentality is necessitated by purely physical phenomena. So by adding (ii) Melnyk comes closer, and perhaps just close enough, to capturing the content of physicalism.[8]

However, while condition (ii) does seem to be an essential part of physicalism, it expresses a type of *supervenience* relation, one that goes beyond the notion of realization itself.[9] One can consistently hold that mentality is realized, and that it is realized at the physical level—in the sense of (a), (b), or R_M—without believing that the mental facts are necessitated by the physical facts as expressed by (ii). Indeed, as mentioned above, by allowing that physical items are realized mentally, R_M is compatible even with mental features being the ultimate realizers. Of course, the fact that (ii) goes beyond the core notion of realization certainly does not prevent one from adding (ii) to the claim that mentality has physical realizers in an effort to better capture the content of physicalism. Still, the notion of realization itself brings us no closer to knowing why we should reject ***II*** (and therefore no closer to knowing why we should accept (ii)). The worry remains that if mental properties are not physical properties, then it's not clear why we should deny the possibility of Absent mentality cases. If mental properties are not physical, then it really does seem that there's a possible world with all the physical facts that actually obtain, including all the same physical laws, but absent some actual mental phenomena. So while Melnyk's condition (ii) might be enough along with realization claim (i) to secure the truth of physicalism, it is not clear why one should accept (ii), i.e., the denial of ***II***, if one believes that mental properties are not physical. So far it appears that the notion of realization, which we hoped would reveal why ***II*** is false, fails to do so.

Definitions (a), (b), and R_M are clearly *functional* accounts of realization, according to which being a realizer consists in playing the functional role characteristic of the realized property/property-instance. However, there are two influential accounts of realization that do not explicitly or obviously fit the functional model. One might wonder whether either of these prevents the mental from realizing the physical, while perhaps also helping us see why ***II*** is false.

---

[8] Although, we might need to add to the facts about the distribution of physical tokens, facts about the physical *types* to which they belong. At the same time, we may wish to drop the idea that the mental facts are a *logical* consequence of the physical facts, and talk instead of metaphysical or even just physical necessitation. (Recall Sect. 2.3.)

[9] However, see how Melnyk (2003, Chap. 2) contrasts the truth-making condition expressed by (ii) with standard supervenience claims.

3.4 Shoemaker's Subset Account

"[I]t is arguable," Shoemaker claims, that the notion of realization "provides the most revealing characterization of physicalism itself: physicalism, we can say, is the view that all states or properties of things, of whatever kind, are physical or physically realized" (2007, p. 1). The description of realization that Shoemaker provides is fittingly called the "subset" account.

In one of his earlier formulations, Shoemaker proposes that "property X realizes property Y just in case the conditional powers bestowed by Y are a subset of the conditional powers bestowed by X" (2001, p. 78).[10] The conditional powers conferred by the belief that it's raining "include, among countless others, one that can be roughly characterized as *being such that if one wants to keep dry and believes that umbrellas keep off rain, this will result in one's taking an umbrella if one goes out*" (2001, p. 82). Suppose that this conditional power, and all of the others conferred at the time by the belief that it's raining, are a subset of those conditional powers conferred to the individual by her having some physical property P. Then, on Shoemaker's subset account, P realizes the belief that it is raining.

The conditional powers of a property are a matter of its potential effects. However, the causal profile of a property includes the potential causes of the property as well as its potential effects. Thus, Shoemaker later expresses his subset account in terms of *forward-looking* and *backward-looking* causal features. He proposes that "one property realizes another if the set of its forward-looking causal features contains as a subset the forward-looking causal features of the other, and the set of its backward-looking causal features is a subset of the backward-looking causal features of the other" (2003, p. 3).

Shoemaker's original subset account was developed as an account of *same-subject* realization, which he labels 'realization$_1$', "in which the instantiation of a property in a thing is realized by the instantiation of another property in that same thing" (2007, pp. 28–29). He recognizes, however, that one might prefer an account

---

[10] Shoemaker also requires that "X is not a conjunctive property having Y as a conjunct" (2001, p. 78). Without this constraint, realization would be too easy to come by. For any property F and any other property G, the conjunction F & G would realize both F and G. However, Shoemaker later comes to realize that a *total* ban on conjunctive properties as realizers is too strict (2007, 26–28).

[11] This different-subject realization is what Gillett (2003) calls "dimensioned" realization.

[12] A reason for denying that they are identical is that they seem to have different modal properties (e.g., the body/lump of clay could survive the loss of personhood/statuehood). Shoemaker also mentions potential differences in historical properties, such as the lump of clay predating the statue (2007, p. 29).

[13] Shoemaker notes, however, that strictly speaking "the realizer in the case of property realization is the instantiation of a property" (2007, p. 3), which arguably is more in line with the idea that to realize is to "make real." He also contends, what philosophers have in mind when they talk of property realization is "the realization of an instance of one property by an instance of another property" (2007, p. 31). So, like Melnyk, Shoemaker thinks of both the realized and the realizer as instances of properties rather than properties themselves.

of realization that allows the bearer of the realized properties to be something other than the bearer of the realizing properties.[11] Shoemaker actually does prefer a different-subject account of realization since he is attracted to the idea that constitution is not identity, and in particular to the view that the person who bears a mental property is not identical with the constituent body (just as the statue is not identical with the constituent lump of clay).[12] The brand of realization he calls 'realization$_2$' obtains when "the properties in one of a pair of coincident entities can be said to realize properties of the other" (2007, p. 29).[13]

Shoemaker defines this second brand of realization, *realization$_2$*, in terms of realization$_1$. Suppose that x belongs to the sortal, *body*, and y is a member of the sortal, *person*. Suppose, also, that body x constitutes person y, and that x has physical feature P and y has mental property M. Then x's having P realizes$_2$ y having M just in case physical property P *plus* the property of being a person realizes$_1$ mental property M, i.e., the set of forward-looking causal features of P *and* being a person contains as a subset the forward-looking causal features of M, and the set of backward-looking causal features of P *and* being a person is a subset of the backward-looking causal features of M. As Shoemaker illustrates: "the instantiation of C-fiber stimulation in my body realizes$_2$ pain in me because (a) I am coincident with my body, (b) I satisfy the sortal *person*, and (c) the conjunction of C-fiber stimulation and being a person realizes$_1$ the property of being in pain" (p. 30). Since realization$_2$ is defined in terms of realization$_1$ and Shoemaker's description of realization$_1$ is a subset account, his description of realization$_2$ also qualifies as a subset account. Although, in what follows, by 'subset account' I indiscriminately refer to both realization$_1$ and realization$_2$.[14,15]

The subset account appears to differ from what Shoemaker calls the standard "higher-order property" account of realization which portrays realized properties as second-order properties the possession of which consists in some first-order property playing the right functional role. Yet, as Shoemaker recognizes, the subset account of realization can be construed as a version of the higher-order property view:

> it can be expressed by saying that the realized property is a second-order property the having of which consists in having some first-order property or other that satisfies a certain condition, namely that its forward-looking causal features include as subset those of the realized property, and its backward-looking causal features are a subset of those of the realized property (2007, pp. 14–15).

---

[14] I refer the reader to Melnyk's (2010) excellent presentation and critique of Shoemaker's realization$_2$.

[15] As mentioned in footnote 11, since realization$_2$ is different-subject realization, it is what Gillett calls the "dimensioned" brand of realization. However, Shoemaker mentions Gillett's notion of the dimensioned account only when he introduces realization$_3$, in which the realizer properties are had by the microconstituents of the properties realized. With realization$_3$, a property instantiation of some macroscopic object is realized "by a microphysical state of affairs involving the instantiation of micro-properties in micro-entities" (2007, p. 32). See 2007, Chap. 3 for details about this "microrealization." I will not discuss Shoemaker's microrealization here, but only mention that the objections I raise to the other two brands of realization apply to this one as well.

Let's focus here, not on the reference to first-order and second-order properties, but on the idea that realization consists in the satisfaction of a certain condition (the playing of a certain role) characteristic of the realized property. If the subset account can be construed in this way, then it should not be surprising that it, like those accounts mentioned in Sects. 3.1–3.3, yields the result that mental properties can realize physical properties. The forward-looking causal features (e.g., the potential behavioral consequences) of *perceiving an approaching lion* or *believing that one is being stalked by a killer* seem to be at least as numerous as the forward-looking causal features of *occupying space* or *having density* or various other general physical properties. The causal powers of the former properties seem at least as numerous as those of the latter—especially since (assuming substance dualism is false) exemplifying the former guarantees exemplifying the latter, but not vice versa. Also, the backward-looking causal features of *perceiving an approaching lion* or *believing that one is being stalked by a killer* would seem to be a subset of the backward-looking causal features of *occupying space* or *having density* (since the potential causes of the latter are far greater). It would seem, then, that in these cases, mental properties realize physical properties on Shoemaker's subset account. So as with (a), (b), and R_M, there is the worry here as well that the appeal to realization does not capture the idea that the mental phenomena obtain in virtue of physical phenomena.

However, it is arguable that this particular worry is avoided with the subset account. For it would seem that any highly specific physical property which, unlike the general physical properties mentioned above, might be sufficient to yield a mental property, does not itself have a mental realizer on the subset account. Take any mental property, M, and any specific physical property, P, that might qualify as a supervenience base for M. It's arguable that (i) the causal powers of P are more numerous than those of M. It's arguable also that (ii) there is no other mental property whose causal powers include as a subset the causal powers of P. Granted, it's not entirely clear that (i) and (ii) are true. But if there is reason to think (i) and (ii) are true, then there is reason to think that no mental property realizes P on the subset account. And if so, then it's arguable that with the subset account there is an asymmetry between mental realizers and physical realizers that we do not find with (a), (b), and R_M, an asymmetry which reflects the fact that the mental is grounded in the physical.

Suppose, then, that we were to accept Shoemaker's account of physical realization. Would we then have a reason to think that ***II*** is false? Would we then know how it could be that the physical facts necessitate the mental facts even though mental properties are not physical? Well, that depends on whether we endorse what Shoemaker calls the *causal theory of properties* (CTP). According to CTP, "for each property there is a causal profile that it has in every possible world in which it can be instantiated, and which is such that having that causal profile is sufficient for being that property" (2007, p. 5). If we add CTP to the view that mental properties are realized in Shoemaker-fashion by physical properties, then we can offer a reason to think that ***II*** is false. For simplicity, consider Shoemaker's realization$_1$ (same-subject, i.e., non-dimensioned/"flat" realization), and suppose that physical property P realizes$_1$ mental property M. Then the forward-looking causal features of M are a subset of those of P, and the backward-looking causal features of P are a subset of those of M. Being facts about causal features, these subset facts will remain the same so long as the causal laws remain the same. Thus, it is nomologically necessary that if P realizes$_1$ M, then whenever P is instantiated M is realized. Now, if the causal profile of a property is essential to that property, as CTP tells us, then it also follows that in any possible world, whenever P is instantiated there is a realization of M; i.e., there is no possible world where something has P in the absence of M. So given CTP and the fact that mentality is realized as the subset account describes, we should expect that the physical facts necessitate (metaphysically necessitate) the mental facts, and we should expect that this is the case even if mental properties are not physical.

Shoemaker does endorse CTP.[16] But for the sake of his 2007 discussion he claims only to presuppose the weaker thesis that each property of a concrete thing "is individuated by a causal profile in the sense that it and it alone has that profile in the actual world and worlds nomologically like it" (p. 142). However, with this weaker version of the causal theory, call it "CTP-" even if the mental is realized by the physical as described by the subset approach it does not follow that the physical facts necessitate the mental facts in the way that physicalism requires. If properties have the same causal profile in all nomologically possible worlds, then what does follow is that all *nomologically* possible worlds with the same physical facts have the same mental facts. But CTP- does not guarantee that *all* possible worlds with the same physical facts (including physical laws) have the same mental facts. CTP- allows, in particular, that a complete physical duplicate of the actual world lacks some of the actual mental episodes. As McLaughlin (2009) points out, with only CTP-, the subset account allows possible worlds at which P (which actually realizes M) is instantiated without any instances of M.[17]

Suppose, then, that we insist on the stronger CTP. As described above, CTP together with the subset account explains why it should be that physical phenomena necessitate mental phenomena even though mental properties are not physical. However, this necessitation is ensured by CTP only at a huge price—the price of requiring much more than what the thesis of physicalism itself entails. The physicalists' belief that mental phenomena obtain solely in virtue of physical phenomena does not require that for each property there is a causal profile that it has in every possible world in which it can be instantiated. One might believe that mentality is solely a function of physical facts (and even believe that mental properties are identical with physical properties) without believing that the causal laws that actually obtain are those that obtain in all metaphysically possible worlds.

---

[16] And he has for some time. See, for example, Shoemaker (1980).

[17] As McLaughlin (2009) notes, Shoemaker recognizes this worry and tries to avoid the problem "by including in the realizer the obtaining of a set of causal laws—normally the laws that obtain in the actual world," in which case, when "the instantiation of property P is said to realize the instantiation of property Q, the full realizer is the occurrence of P together with the obtaining of the laws that give P the causal profile it has in the world in question" (Shoemaker 2007, p. 6). However, if mental properties are not physical, then the laws of nature will include more than just purely physical laws; there will also be irreducibly psychophysical laws, and as McLaughlin indicates, the fact that P together with the psychophysical laws necessitates Q is less than what physicalism demands.

So one can believe that mentality is solely a function of physical facts without believing that properties are individuated in terms of their causal features. In fact, one might believe that the mental is solely dependent on the physical while being quite skeptical about the status of causal laws.

So, while the subset account does seem to provide proponents of CTP with a reason for thinking that ***II*** is false, it does not provide a reason for those physicalists, many in number, who reject CTP. Thus, even if mentality is realized physically in the subset fashion, we are still left with the following worry:

(***II****) unless CTP is true, mental properties not being identical with physical properties prevents the physical facts from necessitating the mental facts in the way that physicalism requires.

This is a worry for all non-reductive physicalists who reject CTP, even if they accept Shoemaker's views on realization.

3.5 Pereboom and Constitution

The lump of clay existed prior to being molded into a statue, and it continues to exist even when remolded and returned to a statueless mass. Since the statue and the lump of clay do not exist at all of the same times, it seems they are not identical. But suppose the lump of clay and the statue miraculously come into existence at the exact same moment and are later annihilated simultaneously. They differ even then. They differ in some of their modal properties; e.g., the lump could have existed prior to the statue and could have remained after the statue was destroyed. So given the necessity of identity, one might conclude that the statue is not identical with the constituent lump of clay. Those who conclude from arguments of this sort that constitution is not identity owe some account of the nature of this peculiar constitution relation ("peculiar" because it seems so much like identity while supposedly falling short).

Baker (2000) offers a rigorous analysis of the constitution relation, an account meant to apply not only to her main concern (the relation between persons and their bodies) but to constitution generally, including the relation between statue and clay, dollar and paper, and flag and cloth. Pereboom also offers a precise analysis of constitution, which focuses on what he calls "material constitution." Since he discusses constitution in the context of defining physicalism, I shall focus on his analysis here.

Pereboom characterizes physicalism as the view that all concrete phenomena are materially constituted, where

x *materially constitutes* y at t if and only if

(a) y is made up of and materially coincident with x at t;
(b) necessarily, if x exists and is in D at t, then y exists at t and is made up of and materially coincident with x at t; and

(c) possibly, y exists at t and it is not the case that y is made up of and materially coincident with x in D at t. (2011, p. 140)

Actually, this is the second of two definitions of material constitution he provides. In the first definition (2011, p. 139), conditions (b) and (c) make no mention of x's being in condition D. However, a reason to favor reference to D is that it captures Baker's insight that in many cases, an F constitutes a G only in a certain context, i.e., only with the right relational features. As Baker illustrates, "it is in virtue of certain legal conventions that a particular piece of paper constitutes a marriage license; it is in virtue of the arrangement of molecules that something constitutes a block of ice; it is in virtue of its evolutionary history that a particular conglomerate of cells constitutes a human heart" (2000, p. 41). Likewise, on a standard functionalist account of mentality, a neural event constitutes a mental episode only by playing the right causal role with respect to sensory input, behavioral output, and other mental episodes. And, as Pereboom describes, on an *externalist* view "the existence of a token belief with some specific content will not be necessitated by the existence of its neural or microphysical constitutor, for in an alternative physical and social environment, this same neural or microphysical constitutor would not yield a belief with that content" (2011, p. 140).

Pereboom's definition of material constitution does not require that an item is identical with what materially constitutes it. One can hold with his account that the statue is materially constituted by the lump of clay while denying that the statue is identical with the lump of clay, and one can accept that a person is materially constituted by her body while denying that she is identical with her body. With this definition, one can also hold that mental processes are not identical with the neural episodes that materially constitute them. In fact, Pereboom (2011, p. 132) argues that just as considerations of multiple realizability have been used to show that mental properties/types are not identical with neural properties/types, multiple realizability can also be used to show that mental *tokens* are not identical with the corresponding neural tokens. Suppose that some mental property M can be realized by tokens of different physical types on different occasions, and suppose that M is realized on some occasion by a token x of some physical type P. Given the multiple realizability of M, M *could have been* realized by a token z of some physical type other than P. So it seems, in this case, that while token y of M is accompanied by physical token x, it is accompanied by x only contingently. Given the necessity of identity, Pereboom concludes that x ≠ y.

This reasoning can easily be resisted. It is arguable that while any given mental property could be instantiated in tokens of different physical types, no one token of the mental type could *itself* have been constituted by any physical token other than

[18] Baker (2013, pp. 739–740) makes the point that the *multiple realizability* that condition (c) entails should not be required in an account of constitution, noting that if some diamond were realized by a different lattice of carbon atoms, it would arguably not be the same (numerically identical) diamond. Her conclusion regarding Pereboom's account is a bit different from mine. She is making a claim about what *constitution* (not physicalism) should allow, and she is certainly not

the one that actually did occur on that occasion. It is arguable that if M were realized on that occasion by a token of some physical type other than P, then while M would still have been instantiated, the particular token of M that actually did occur on that occasion would have failed to occur. Still, while Pereboom's argument may be resisted, it is not implausible to think that a definition of physicalism should allow a physicalist to believe that mental tokens are constituted and not identical with the corresponding physical tokens. (Although, the definition should certainly also allow a physicalist to endorse token-identity, which suggests that condition (c) should not be included in an account of constitution fit for defining physicalism.[18])

In conditions (b) and (c) of his definition, 'D' serves to indicate whatever contextual/relational conditions are required to have the constituted token. And with the idea of a description 'D' whose satisfaction makes it the case that constitution obtains, Pereboom has described what may be considered genuine *realization*, where 'D' indicates the functional role to be played for realization to obtain. But remember the issue described in the earlier sections of this chapter regarding mental phenomena realizing physical phenomena. Since Pereboom's notion of constitution may be viewed as a brand of realization, the question arises whether the mental can constitute the physical on his account. Actually, his account seems to avoid the worry that the mental can realize (i.e., constitute) the physical. According to condition (a), "y is made up of and materially coincident with x." The requirement that the constituted is *made up of* what constitutes it seems to prevent higher-level (e.g., mental) items from constituting items at a lower (physical) level. For it would seem that what's at the lower-level always *makes up* higher-level phenomena, and never the other way around. As he says, "the *made up of* relation is asymmetric… the lattice is not made up of the diamond" (p. 138). So with the asymmetric nature of the *made up of* relation, it seems we avoid the implausible view that the mental constitutes the physical. In this respect at least, Pereboom's account is better suited to defining physicalism than the more paradigmatic approach to realization discussed in Sects. 3.1–3.3. Unfortunately, his constitution view still falls short of capturing all that physicalism requires.

He mentions that by 'made up of' he means '*wholly* made up of'; to say that a collection x of physical processes makes up a mental episode y is meant to entail (as physicalism requires) that y is not made up of some non-physical processes as well. However, Pereboom offers little to explain what exactly the (wholly) *made up of* relation amounts to. He admits that "the core of the *made up of* relation is unanalyzable and thus primitive," although he resists "the claim that this proposal amounts to obscurantism," since it "is sufficiently clear what we mean when we say that the diamond is made up of a lattice of carbon atoms, and that the brain is made up of a configuration of various kinds of neurons, even if no reductive analysis is provided

(Footnote 18 continued)
claiming that constitution should allow *token-identity*. Yet, what she says about the diamond does serve to illustrate the point I'm making here that Pereboom's argument from multiple realizability to token non-identity can be resisted.

for this relation" (p. 138). However, even on an intuitive understanding of the made up of relation, it is clear that his constitution account does not capture all that physicalism demands. Even with a rough, intuitive grasp of what it is to be made up of, it is clear that an item y can be made up of (wholly made up of) physical items even though y's having the mental properties it has is not solely a matter of the physical details. A non-physicalist can believe that a mental episode is made up of chemical processes, which are wholly made up of the stuff of physics, without believing that the mental properties of the episode are wholly dependent on purely physical features. It is true that to remain faithful to physicalism, the non-reductionist must insist that all mental tokens are either identical with, or decomposable into, or made up of physical tokens. But this constitution claim is not enough to make one a physicalist, since one can accept the constitution claim, and even the token-identity claim, while denying that mental properties obtain solely in virtue of physical properties. One can believe that a mental token is wholly made up of physical tokens while also believing that what gives the mental token its mental status (what makes it belong to a certain mental type) are its irreducibly subjective, non-physical features. That's why rejecting *substance dualism* does not guarantee that one is a physicalist, which is why many property dualists can and do reject physicalism.[19]

It follows that (a) of Pereboom's analysis does not itself secure the content of physicalism; although, one might think that with (b) and (c), he adds enough to secure all that physicalism requires. But that is not the case. According to condition (b), *necessarily, if x exists and is in D at t, then y exists at t and is made up of and materially coincident with x at t.* It is not required that condition D, which x must satisfy to constitute y, is a purely physical condition. So, for all that (b) demands, it might be that physical token x constitutes mental token y only by exemplifying some irreducible element of subjectivity or qualitative character. One step toward precluding this sort of case is to insist that *D is purely physical.* To capture the content of physicalism, we will also need to require that the necessity mentioned in (b) is stronger than nomic necessity; otherwise, the condition could be satisfied with the help of irreducibly psychophysical laws, which would not guarantee that token y is present solely in virtue of the physical facts.[20]

Pereboom (2013) admits that to secure the necessitation essential to physicalism his definition would "require supplementation by a no-emergent-law condition" (p. 762). Or, instead, we might specify that the necessity mentioned in (b) is *physical necessity.* Given that condition D is a purely physical condition, then we are close to capturing the idea that the physical facts (physical properties/relations

[19] By 'property dualist' I mean anyone who denies that mental properties are physical properties, which includes all non-reductionists, even those who support physicalism.

[20] In response to Pereboom's (2002) claim that the physical token together with the requisite relational features are sufficient for the mental token, Melnyk points out, "for all that Pereboom says, this sufficiency might be sufficiency in accordance with a fundamental law of physical-to-mental emergence" (2008, p. 1292).

instantiated and the physical laws) necessitate the mental facts. Although, to ensure this result we should also say something about the types, the properties, that x and y instantiate. We want to be assured that x yields y in virtue of the physical properties that x instantiates, physical properties that might extend beyond the specified condition D; and, of course, we also want to be assured that x's being of that physical type (under those conditions) necessitates that y is of that same mental type. With all of the qualifications in place, the definition becomes something like: Token x of physical type P *materially constitutes* token y of mental type M at t if and only if

(a) y is made up of and is materially coincident with x at t, and
(b) it is physically necessary that if x exists at t and is of type P in physical condition D, then y exists at t and is of type M, and is made up of and materially coincident with x at t.[21]

But now condition (b) contains a supervenience thesis of brand N_P, which tells us that the physical details together with the purely physical laws necessitate the mental details. This is a condition that physicalism seems to entail.[22] And with this supervenience condition as part of our definition (as it seems it should be if used to formulate physicalism), we are left with our nagging question of why the physical facts should necessitate the mental facts in this way if mental properties are not identical with physical properties. The notion of being *made up of*, for all that Pereboom says about it, fails to answer this question; as mentioned earlier, one can believe that a mental token is wholly made up of physical tokens while also believing that what gives the mental token its mental status are its irreducibly subjective, non-physical features.

It seems, then, that as it stands Pereboom's constitution analysis falls short of capturing the content of physicalism. And if the analysis were strengthened (along the lines above) to entail all that physicalism requires, we would still be left with worry ***II***. This issue, again, is not whether there is reason to accept the physical necessitation that physicalism requires; the issue is how this necessitation could even obtain if mental properties are not physical. For if we concede that mental properties are not physical, then we seem compelled to allow possible worlds that physically duplicate the actual world (same physical laws and same distribution of physical properties/relations) but without some of the mental episodes that actually obtain.

---

[21] Condition (c) is deleted since, as was mentioned, while the denial of token-identity perhaps should be allowed, it should not be required by a definition of material constitution fit for defining physicalism.

[22] Physicalism entails the physical necessitation of N_P, or at least N_P* (as described in § 2.3). Recall that N_P* is the physical necessitation claim restricted to *positive* mental facts (to allow the possibility of mental extras).

3.6 Recap

Mentality is realized by physical phenomena. The basic and standard idea here is that a mental property is a functional property, which is instantiated by some physical process in virtue of the physical process playing the functional role definitive of the mental property. We have seen that this basic notion of realization allows not only that the physical realizes the mental, but also that the mental can realize the physical; and this is possible even with Melnyk's rigorous definition of realization.

The fact that there are cases in which physical properties are realized by mental items would seem to undermine the prospect of characterizing physicalism in terms of realization. For if a mental item can realize a physical property, then it seems that whatever makes it the case that mentality obtains in virtue of physical phenomena, it is not simply that the former is realized by the latter. If it were true that there is some level of physical structure at which none of the physical realizers of mentality themselves have mental realizers, then perhaps the notion of realization could be used to at least capture the idea that the physical is primary. However, as shown in Sects. 3.2 and 3.3, even fundamental physical properties have mental realizers (at least where realization is construed in terms of (a), (b), or Melnyk's R_M). The fact that even basic physical properties are realized by mental items certainly threatens the prospect of characterizing physicalism in terms of realization.

We have also seen that by allowing the mental to realize the physical, the notion of realization on the standard construal (specified in terms of (a), (b), or R_M) does not help to show why the worrisome ***II*** is false. We are still left wondering how it could be that the mental facts are necessitated by physical facts (as physicalism requires) if mental properties are not physical. In fact, since the mental can realize the physical, and even the stuff of fundamental physics, the idea that mentality is realized by physical items is compatible with the view that all of the ultimate realizers are mental instead of physical, and therefore that the mental necessitates the physical rather than the other way around.

One non-standard view of realization is Shoemaker's subset account (his realization$_1$ and realization$_2$). The subset approach does help us see why ***II*** might be false—but only given CTP, according to which, each property has a causal profile and the same causal profile in every possible world. So while the subset account provides proponents of CTP with a reason for thinking that ***II*** is false, it does not offer a reason for the many non-reductive physicalists who reject CTP.

Another non-standard view of realization is provided by Pereboom with his description of material constitution. Since Pereboom's account relies on the notion of being *made up of*, with his account it seems we avoid the result that the physical is

[23] Another non-standard view of realization is provided by Yablo (1992), who describes the relation between mental properties and their physical realizers as a relation of *determinables* to *determinates*. In Francescotti (2010), I show that Yablo's account also has the worrisome consequence that the mental can realize the physical.

realized by (constituted by) the mental; if physical token x is materially coincident with y and y is made up of x, then given the asymmetry of the *made up of* relation, x is not made up of y. However, we have seen that as it stands Pereboom's account still falls short of capturing the content of physicalism. And if the account were strengthened to entail the sort of physical necessitation that physicalism requires, then we would be left again with worry ***II***, wondering how it could be that the physical necessitates the mental in the way physicalism requires if mental properties are not physical. The notion of being "made up of" does nothing to answer this question.[23]

In the next chapter, I offer a definition of physicalism that I think does reveal why ***II*** is false, showing that the fact (if it is a fact) that mental properties are not physical does not preclude the necessitation of the mental by the physical that physicalism demands. The account offered is a way to make sense of the idea that mentality is constituted by physical phenomena, and one which comes closer than Pereboom's to capturing the idea that the former *depends wholly on*, *consisting in nothing other than*, the latter. The account also borrows from Shoemaker's idea that there is something about mentality that is a *subset* of what obtains physically—not a subset of *causal powers*, but a subset of the *concrete items* involved. I warn, however, that while the definition of physicalism provided in Chap. 4 shows why ***II*** is false, thereby showing how a non-reductionist can earn the title 'physicalist', it does so only by suggesting a couple of plausible and robust definitions of 'physical property' according to which mental properties are in fact physical (even if they are not identical with any properties of the natural sciences).

References

Baker, L. 2000. *Persons and bodies: A constitution view*. Cambridge: Cambridge University Press.

Baker, L. 2013. Pereboom's robust nonreductive physicalism. *Philosophy and Phenomenological Research* 86: 736–744.

Block, N. 1978. Troubles with functionalism. In *Minnesota studies in the philosophy of science*, vol. 9, ed. C.W. Savage, 261–325. Minneapolis: University of Minnesota Press.

Block, N. 1990. Can the mind change the world? In *Meaning and method: Essays in honor of Hilary Putnam*, ed. G. Boolos, 137–170. Cambridge: Cambridge University Press.

Francescotti, R. 2010. Realization and physicalism. *Philosophical Psychology* 23: 601–616.

Gillett, C. 2003. The metaphysics of realization, multiple realizability, and the special sciences. *The Journal of Philosophy* 100: 591–603.

McLaughlin, B. 2009. Sydney Shoemaker, *Physical realization. Notre Dame Philosophical Reviews*.

Melnyk, A. 1996. Formulating physicalism: Two suggestions. *Synthese* 105: 381–407.

Melnyk, A. 2003. *A physicalist manifesto*. Cambridge: Cambridge University Press.

Melnyk, A. 2006. Realization and the formulation of physicalism. *Philosophical Studies* 131: 127–155.

Melnyk, A. 2008. Can physicalism be non-reductive? *Philosophy Compass* 3: 1281–1296.

Melnyk, A. 2010. Comments on Sydney Shoemaker's *Physical realization. Philosophical Studies* 148: 113–123.

Papineau, D. 1993. *Philosophical naturalism*. Oxford: Blackwell.

Pereboom, D. 2002. Robust nonreductive materialism. *The Journal of Philosophy* 99: 499–531.

Pereboom, D. 2011. *Consciousness and the prospects of physicalism*. New York: Oxford University Press.

Pereboom, D. 2013. Replies to Daniel Stoljar, Robert Adams, and Lynne Baker. *Philosophy and Phenomenological Research* 86: 753–764.

Polger, T., and L. Shapiro. 2008. Understanding the dimensions of realization. *The Journal of Philosophy* 105: 213–222.

Shoemaker, S. 1980. Causality and properties. In *Time and cause*, ed. P. van Inwagen, 109–135. Dordrecht: D. Reidel.

Shoemaker, S. 2001. Realization and mental causation. In *Physicalism and its discontents*, ed. C. Gillett, and B. Loewer, 74–98. Cambridge: Cambridge University Press.

Shoemaker, S. 2003. Realization, micro-realization, and coincidence. *Philosophy and Phenomenological Research* 67: 1–23.

Shoemaker, S. 2007. *Physical realization*. New York: Oxford University Press.

Van Gulick, R. 1992. Nonreductive materialism and the nature of intertheoretical constraint. In *Emergence or reduction: Essays on the prospects of nonreductive physicalism*, ed. A. Beckermann, H. Flohr, and J. Kim, 157–179. Berlin: Walter de Gruyter.

Yablo, S. 1992. Mental causation. *The Philosophical Review* 101: 245–280.

Chapter 4
Defining Physicalism

4.1 Token-Identity and Being Nothing More Than

The thought that mental properties are multiply realizable is the main reason why type-identity theories began to go out of favor in the 1960s and soon became and remained the minority view. Yet, many who rejected psychophysical type-identity claims still endorsed the weaker *token*-identity claim that every token of a mental type is a token of some physical type or other, though not necessarily the same physical type on each occasion. Expressed in terms of properties and their instances, the view is that

(TKI) For any mental property M, and any instance x of M, there is a physical property or collection of physical properties, P, and an instance y of P such that x = y.

It is widely recognized that endorsing this token-identity claim is not by itself enough to make one a physicalist. Even if an instance of some mental property M is also an instance of a physical property P, it might be that what makes it an instance of M is something more than its being an instance of P or any other physical property. Thus, one can consistently accept TKI while also rejecting physicalism, by claiming (for example) that the physical process that instantiates M manages to do so only because of certain irreducibly subjective features, ones that do not depend solely on the presence of any set of physical features. So TKI does not entail physicalism regarding mentality. However, as will now be shown, there are derivatives of TKI that do capture the physicalist belief that mental facts obtain solely in virtue of physical facts.

Suppose that I engage in some peculiar body movements on some occasion and in doing so I thereby engage in a (pathetic) display of dancing. The property of being engaged in dancing is not the same as the property of moving one's body about as I did on that occasion, since dancing can and (thankfully) often does occur without those peculiar movements I made. Still, it does seem that I have engaged in

R. Francescotti, *Physicalism and the Mind*,
SpringerBriefs in Philosophy, DOI 10.1007/978-94-017-9451-0_4

dancing on that occasion in virtue of nothing more than moving my body in just those ways at that time with the intention to dance.

In this case, the properties instantiated differ. There is the property, D, of *being engaged in dancing*. There is also the property, B, of *intentionally moving one's body in those dance-resembling ways*. And property D is not the same property as B since dancing can and often does occur without those particular movements I made on that occasion. However, despite the difference in properties, the act that instantiates D in this case is the very same act as the one that instantiates B. We are imagining that the very series of movements that is my act of dancing on that occasion is the moving of my body in just those ways. So the x that is an instance of D is an instance of B.[1] But that's not all. It also seems that x's instantiation of D on this occasion is *nothing more than* x's instantiation of B. That very process of dancing seems to be nothing more than that particular series of dancing-intention driven body movements on that occasion.[2]

So, in this case, there are 3 pairs of items to consider. There are the properties, D and B, and in this case,

(i) property D $\neq$ property B.

Also, there is the event or process, x, that instantiates D, and the y that instantiates B, and in this case it seems that x is the same as y. That is, the very process that is an instance of moving one's body in those ways is also an instance of dancing. So in this case,

(ii) there is an instance x of D and an instance y of B, and x = y.[3]

Furthermore, there is *x's instantiation of D* and *y's instantiation of B*, and it seems that:

(iii) the instantiation of D on that occasion is nothing more than the instantiation of B.

To make the difference between (ii) and (iii) clearer, consider a property more generic than B. Consider the property of *moving one's body*. Call it "B-." Now in the case we are imagining, the activity x, which is an instance of B and D, is also an instance of property B-; an instance of moving one's body in that specific dancing manner requires an instance of the simpler property of moving one's body. However, it seems quite clear that x's instantiating D is something more than x's instantiating B-. What makes x an act of dancing is not simply that it is an act of moving one's body. What makes it an act of dancing is that it is an act of moving one's body with the right intentions in certain specific ways. In general, then, there

[1] Or I should say, this is what seems to be the case. Thoughts of *constitution* and its difference from identity (which will be addressed shortly) might suggest otherwise.

[2] I specify that the movements are driven by dancing intentions to avoid the potential objection that being engaged in genuine dancing requires intending to dance (as opposed to, say, moving in that way due to a seizure).

[3] Again, issues of constitution versus identity are put aside for the moment.

is a difference between the same process instantiating two different properties, F and G, and the instantiation of F in that case being nothing more than the instantiation of G. In some cases, there is the former without the latter.

To take a more famous example, consider the stabbing of Caesar. The property of stabbing is different from the property of killing since one can kill in many ways other than stabbing. However, Brutus and the others performed an action, x, that was an instance of both properties; it was a collective stabbing and it was a killing. So the killing/stabbing counterpart to (ii) seems to be true. Moreover, the analogue of (iii) also seems to be true. Philosophers debate whether the killing of Caesar is the *same event* as the stabbing of Caesar (a debate whose answer depends on whether events are individuated in terms of the properties exemplified). Yet, regardless of one's views on events, it does seem that in some quite plausible sense of 'nothing more than', the killing of Caesar on that occasion was nothing more than the stabbing of Caesar in exactly that manner (a manner of stabbing which includes being death-producing).

Now let's return to mental properties. Suppose that some mental property M is not identical with the underlying physical property P. Also suppose that while the type-identity thesis is false, the token-identity claim is true and

(ii_M) there is an instance x of M and an instance y of P, and x = y.

Even assuming (ii_M) is true, there is the further question of whether

(iii_M) the instantiation of M on that occasion is nothing more than the instantiation of P.

There is a difference between (ii_M) and (iii_M) since one can accept the former while rejecting the latter. One might think that while an instance x of M is an instance of some physical property, what makes x an instance of M is a matter of more than its physical features. One can accept token-identity claims of type (ii_M) while agreeing, for example, with Jackson's (1982, 1986) classic view that the qualitative character of a mental state is not exhausted by its physical features.[4] And one can accept (ii_M) while endorsing Chalmers' (1996) proposal that phenomenal properties are fundamental features of the world, or even his suggestion that both phenomenal and physical features ultimately depend on proto-phenomenal properties. Clearly, then, a non-physicalist can consistently accept (ii_M). However, with the "nothing more than" claim, (iii_M) seems to be something that only a physicalist can consistently accept. For the claim that an instantiation of M on some occasion is nothing more than the instantiation of P rules out the presence of M on that occasion being due to something other than the presence of physical features. And the instantiation of M being a function of none other than the instantiation of physical features is just what, and apparently only what, a physicalist regarding M would claim. So, as an initial attempt at defining physicalism regarding mentality, we might try

---

4 The non-physicalist property dualism Jackson defends there is a view he has since rejected. See, e.g., Jackson (2003, 2012).

(NMT) For any mental property M, and any instance x of M at a time t, there is some physical property or collection of physical properties, P, such that x instantiates P at t, and x's instantiation of M at t is *nothing more than* x's instantiation of P at t,

where 'NMT' abbreviates 'nothing more than'.

To know how well NMT captures the content of physicalism, we obviously need to get clear on what the *nothing more than* relation amounts to. It would be nice, for the sake of simplicity, to regard the nothing more than relation as *identity*. Then NMT would amount to the following special brand of token-identity claim:

(TKI*) For any mental property M, and any instance x of M at a time t, there is some physical property or collection of physical properties, P, such that x instantiates P at t, and x's instantiation of M at t = x's instantiation of P at t.

Unlike the standard token-identity thesis, characterized by TKI, with this restricted token-identity claim what's being identified are not merely tokens of M and P, but tokens of the higher-order properties of *being an instantiation of M* and *being an instantiation of P*. And for this reason, TKI does not entail TKI*. It might be that instance x of M is an instance of P without it being the case that x's instantiation of M is identical with x's instantiation of P. Consider any scenario in which (ii_M) is true and (iii_M) is false, e.g., a case in which x is an instance of both M and P but x's having the qualitative character essential to M is not solely a matter of any of the physical features x instantiates. In a case of this sort, x is an instance of both M and P, but x's instantiation of M is not identical with, since it is not nothing more than, its instantiation of P.

The appeal to identity does add precision to talk of "being nothing more than," but there are reasons *not* to rely on identity in our explication of NMT. While physicalism itself is a metaphysical doctrine, there are a variety of metaphysical issues on which a definition of physicalism should remain neutral. One of these issues is how the property-instantiations I've been referring to are best construed. Should they be considered *events*, or should they be thought of as *states of affairs*, or are they best viewed as belonging to some other ontological category? And whichever of these ontological categories we choose, what exactly is the nature of the items belonging to that category? It seems that a definition of physicalism should remain neutral on such issues. One's being a physicalist does not seem to depend on one's stance regarding these matters. However, whether TKI* is true does depend on the answer to these questions.

Suppose, for example, that we regard the items identified by TKI* as states of affairs. On the standard view, states of affairs are complex items with objects and properties as constituents. The state of affairs of Fido's being fierce, on the standard view, is an ontologically complex item that is comprised of the individual Fido and the property of being fierce. But if the items identified by TKI* are states of affairs construed in this manner, then TKI* cannot be true unless mental properties are identical with physical properties. Or suppose the items that TKI* identifies are to be understood as *events*. Then whether TKI* is true depends in part on the correct view of events. On Kim's (1966) fine-grained account, an event (like a state of affairs) is a structured item, comprised of an object (or objects), a property (or relation), and a time. This view entails that events are identical only if the constituent properties are identical. So if events are what TKI* claims to be identical, then given Kim's account of event-identity, TKI* cannot be true unless mental properties are physical. On the other hand, if events are individuated more loosely, e.g., in terms of their spatiotemporal location, then TKI* could be true even if $M \neq P$.[5]

It seems that whether one deserves the label 'physicalist' should not depend on whether one regards the property-instantiations I mentioned as events or as states of affairs, or as something else altogether, and it seems that whether one qualifies as a physicalist should not depend on how finely or coarsely one chooses to individuate such items. Given that a definition of physicalism should be neutral on these metaphysical matters, we should not include in our definition the identity claim expressed by TKI*.

It's arguable that we should also strive for a definition of physicalism that remains neutral on whether *constitution* is identity. As mentioned in Sect. 3.5, Pereboom (2011, p. 132) points out that just as considerations of multiple realizability have been used to show that mental types are not identical with physical types, they can also be used to show that mental *tokens* are not identical with their physical constituents. Suppose that some mental type M is realized by tokens of different physical types on different occasions, and suppose that M is realized on a certain occasion by a token y of some physical type P. Given the multiple realizability of M, Pereboom reasons that the token x of M on that occasion *could have been* accompanied by a token z of some physical type other than P. And if so, then it seems that while x is in fact accompanied by y, it is so only contingently. Given the necessity of identity, Pereboom concludes that $x \neq y$. I noted in 3.5 that this reasoning can be resisted. It is arguable that while any given mental property could be instantiated by tokens of different physical types, no one token of the mental property could itself have been constituted by any physical token other than the one that actually did occur on that occasion. It is arguable that if M were realized on that occasion by a token of some physical type other than P, then while M would still have been instantiated, the particular token of M that actually did occur on that occasion would have failed to occur.[6] And if so, then we can accept the token-identity claim along with the multiple realizability of mental properties without having to admit that instance x of M is only contingently identical with instance y of P. Still, while Pereboom's argument may be resisted, it is arguable that a definition of physicalism should allow that the constitution of mental tokens by

---

[5] See, for example, Quine (1950, 1985) and Davidson (1969) on events.

[6] As Baker (2013, pp. 739–740) points out in her discussion of Pereboom's account, if some diamond were realized by a different lattice of carbon atoms, it would arguably not be the same (numerically identical) diamond. (Recall footnote 18 of Chap. 3.)

physical tokens is not a matter of genuine identity. It's arguable that nothing about the content of physicalism itself entails that mental tokens are identical with physical tokens; one can arguably remain faithful to physicalism while denying mental-physical token-identity for either Pereboom's reason or some other reason.

So, for the purpose of defining physicalism, perhaps we should prefer NMT with its emphasis on being nothing more than. And to avoid any hint of a token-identity claim, we might wish to rephrase NMT slightly, as follows:

(NMT) For any mental property M, and any instance x of M at a time t, there is some physical property or collection of physical properties, P, and an instance y of P at t, such that x's instantiation of M at t is nothing more than y's instantiation of P at t.

It seems that NMT is sufficiently neutral on issues of constitution and identity, and neutral enough on questions about the nature of the property-instantiations mentioned, to be acceptable to all physicalists. Yet, whether NMT is something that only a physicalist would accept cannot be decided until we clarify the obscure talk of being *nothing more than* to which we have returned.

It might be thought that the notion of *grounding*, which has been the topic of much recent debate in metaphysics, might help clarify matters.

4.2 Grounding

Talk of grounding is meant to capture the intuitive idea of one fact *consisting in* or holding *in virtue of* another fact. Examples of grounding talk include the following statements mentioned by Correia (2010, p. 251).

> Dispositional properties are grounded in categorical properties.
> Legal facts are grounded in non-legal, e.g., social, facts.
> Morally wrong acts are wrong in virtue of non-moral facts.
> Universals exist in virtue of their having exemplifiers.
> Determinables are exemplified in virtue of corresponding determinates being exemplified.
> Holes are grounded in facts about their hosts.

Many who defend the notion of grounding as a valuable philosophical concept accept the grounding relation as primitive and therefore indefinable. Still, advocates of grounding generally are prepared to state what they consider some of its distinctive logical and structural features in an effort to make the notion less than unacceptably obscure.[7]

---

[7] See Rosen's (2010) helpful description of these logical and structural features. The attempt to illuminate the concept of grounding also involves giving examples of its application (including those in the list above) and describing its connection to other philosophical notions, e.g., the notions of truth-making, reduction, and explanation.

One feature often considered crucial to grounding is *determination*, i.e., *necessitation*; it has been claimed that one fact is grounded in another only if the latter necessitates the former. As Rosen puts it, "If [p] is grounded in [q], then q entails p"; and when grounding is understood as a type of metaphysical dependence, the "facts that ground [p] together ensure as a matter of metaphysical necessity that [p] obtains" (2010, p. 118). Insofar as grounding is a necessitation relation, it might be thought that the notion of grounding can be used to capture the idea that the mental *supervenes* on the physical since supervenience itself is a necessitation relation.[8] Another logical feature commonly cited in discussions of grounding is *asymmetry*. It is thought that if one set of facts is grounded in another, then the latter is not grounded in the former. The appeal to asymmetry honors the idea that the ground is *more fundamental* than the grounded features. For this reason, *irreflexivity* has also been considered definitive of grounding since no fact can be more fundamental than itself. *Transitivity* is also mentioned in discussions of grounding. It is tempting to think that if, for example, a moral fact is grounded in a behavioral fact and the behavioral fact is grounded in a microphysical fact, then the moral fact is grounded in the microphysical fact.[9]

Necessitation seems to be a relation a physicalist would think obtains between physical facts and mental facts. However, it is doubtful that the sort of physical dependence that is essential to physicalism has the other logical features mentioned above. It is a matter of some debate whether one can consistently be a physicalist while denying the mental properties are physical; but obviously there is no question that one can be a physicalist while believing that mental properties are physical and therefore that mental facts are physical facts (on a non-conceptual, worldly view of facts). Indeed, the acceptance of psychophysical property- and fact-identities would seem to be physicalism at its most robust. Yet, unlike the grounding relation, identity is reflexive and symmetrical. Thus, a definition of physicalism should not require that the dependence of the mental on the physical is irreflexive and asymmetrical as grounding is often thought to be.

It is also arguable that the necessitation of the mental by the physical is not quite the same sort of necessitation as that found in many of the examples used to illustrate grounding. Granted, the idea that mental facts are grounded in physical facts is sometimes used as an example of the grounding relation; the first example of grounding talk on Correia's (2010) list is 'Mental facts obtain in virtue of neurophysiological facts'. But there seems to be a difference between the necessitation relation in this case and that present in his other examples of grounding. Assuming that mental properties are not identical with the underlying physical properties, an instance of the latter necessitates an instance of the former only with the help of the *physical laws* that obtain; even a physicalist can (and perhaps should) hold that the

---

8 See Correia's (2008, pp. 1027–1029) discussion of supervenience and ontological dependence.

9 Schaffer (2012) offers counter-examples to the transitivity of grounding, and then proposes a *contrastive* approach to grounding along with a *differential transitivity* principle to handle the problem cases.

physical properties instantiated might yield a different distribution of mental properties in a world with different physical laws. Of course, physicalism is often construed as the *metaphysical* necessitation of the mental by the physical, i.e., every possible world that physically duplicates the actual world mentally duplicates it as well. The physical duplication, however, is assumed to include duplication of the physical laws. The idea is that all possible worlds that duplicate the actual world in the distribution of physical properties/relations and the physical laws that obtain are worlds that duplicate the actual world mentally.

The role of physical laws in the necessitation of the mental by the physical seems to make this case different from the other cases cited above from Correia's list. If it is true that *universals exist in virtue of their having exemplifiers*, then it seems that it's true because it is part of the essence of universals that none go unexemplified. The laws of physics are not what make it the case, if it is the case, that no universals go unexemplified. And if it is true that *morally wrong acts are wrong in virtue of non-moral facts* (e.g., the fact that a surplus of suffering is produced), then it seems that that's because of the very nature of morality. While the laws of physics, chemistry, and biology are what make certain stimuli productive of pain and suffering, these laws of nature are not what make the production of suffering wrong. Also, suppose it is true that *determinables are exemplified in virtue of corresponding determinates being exemplified.* Suppose, to use one of Rosen's (2010, p. 126) examples, that the ball is blue in virtue of being cerulean. Of course, the laws of physics determine whether an object is cerulean or some other shade of blue. But the truth of the statement 'If cerulean, then blue' has nothing to do with the physical laws; it is part of the very essence of being cerulean that cerulean objects are blue.[10]

It is doubtful, then, that the necessitation of the mental by the physical is the same sort of necessitation that obtains in most of the typical examples of grounding.[11] And even if I am wrong about this putative difference, one can certainly believe that there is this difference while still deserving the label 'physicalist'. Also, as mentioned above, one can warrant the title 'physicalist' while believing that some of the other logical features used to characterize grounding (e.g., asymmetry and irreflexivity) are not true of the relation between the mental and the physical. Thus, insofar as NMT is being offered as a definition of physicalism, it is perhaps advisable to leave talk of grounding behind here and try to find some other way to clarify the idea of being *nothing more than* at work in NMT.

---

[10] For a discussion of the connection between essence and ground, see, e.g., Fine (2012, §I.11) and Rosen (2010, §13). Also see Fine (1994, 1995) for discussion of essence and ontological dependence in general.

[11] Fine distinguishes between the metaphysical, the normative, and the natural conceptions of ground, and he suggests that assuming mental properties are not physical, "the view that the mental is grounded in the physical is only plausible for the natural conception" (2012, p. 77). (As a non-mental example of a statement of natural ground, he gives "The fact that the particle is accelerating obtains in virtue of the fact that it is being acted upon by some net positive force" 2012, p. 37.) If it is true that the mental is grounded in the physical only on the natural conception of ground, then the type of ground that connects the mental to the physical is not the metaphysical brand that discussions of grounding usually concern.

4.3 The Principle of Concrete Exhaustion

Recall the dancing example from Sect. 4.1. I exhibit some peculiar body movements on some occasion and by doing so I thereby engage in a display of what might be called "dancing." The property, D, of being engaged in dancing is not the same as the property, B, of moving one's body in just those dancing-intention driven ways on that occasion, since dancing can and often does occur without those particular movements I made. Still, it does seem that I am dancing on that occasion in virtue of nothing other than the instantiation of B at that time. So it seems true to say that the instantiation of D that occurred then is nothing more than the instantiation of B on that occasion. However, recall the coarser property, B-, of *moving one's body*. It's clear that what makes my act an instance of dancing is not simply that it's an act of moving my body since there are many series of body movements that are not instances of dancing. So it is not accurate to say that the instantiation of D that occurred on that occasion is nothing more than the instantiation of B-.

But what's the crucial difference here between the instantiation of B and the instantiation of B-? Why does it seem plausible to say that nothing more than the former is involved in the act of dancing, while something more than the latter is involved? Well, strictly speaking, there is something more going on in both cases. Since D $\neq$ B and D $\neq$ B-, an extra *property* is instantiated in both cases. The difference between the two cases has to do with what *concrete* items are involved—where the word 'concrete' here is used to denote any item that is not abstract (and therefore not a property, construed as either a universal or a trope). Let's use the word 'item' loosely enough to include all particulars—not just those one would normally label 'objects', but also the events and processes that occur within those objects and which those objects undergo. Then the point is that the concrete items involved in the instantiation of D on that occasion includes nothing more than whatever concrete items were needed for B to be instantiated at the time. That is, the sum of concrete items at that time that make it the case that D is instantiated does not include any concrete items other than those required for the instantiation of B. However, the concrete items involved in the instantiation of D do include more than those that the instantiation of B- requires. Put simply, dancing involves more, in the realm of the concrete, than simply moving one's body—although if one moves one's body in just that manner (and with those intentions) on that occasion, then enough has obtained for dancing to have thereby occurred.

Brutus and the others stabbed Caesar, and in so doing, they killed Caesar. Philosophers debate whether the killing of Caesar and that series of stabbings qualify as the same event. However, one can remain neutral on how to individuate events while acknowledging that the killing of Caesar was nothing more, in terms of all items concrete, than the stabbing of Caesar in just that way on that occasion. That is, the set of concrete items (objects, events, processes, or whatever else concrete) that made it the case that an act of killing occurred on that occasion contains nothing more than those concrete items at the time required for that death-causing series of stabbings.

Let's use the phrase 'concretely exhausted by' to denote the relation that obtains when an instantiation of one property is nothing more, in terms of everything concrete, than an instantiation of some other property. And as an initial attempt to define this relation, let us start with:

> An instantiation of F on some occasion is *concretely exhausted* by an instantiation of G on that occasion = the set of concrete items on that occasion that makes it the case that F is instantiated includes nothing more than the set of concrete items on that occasion required for the instantiation of G.

To know whether the notion of concrete exhaustion can be used to adequately define physicalism, the analysis needs to be made more precise. Let's leave it open whether the instantiation of D, or B, or B- is best viewed as an event, a state of affairs, an occurrence, or a fact. (And I think this should be left open in a definition of physicalism.) Although, we do need to get clear on what is meant by saying that some set of concrete items *makes it the case* that a property is instantiated, or that it is *required* for the instantiation of some property.

Let's start with 'makes it the case that'. How should that expression be construed here? I think it is not implausible, for the purpose of defining physicalism, to interpret the expression as denoting a type of sufficiency, and a relatively strong brand of sufficiency. Conceptual sufficiency seems to be the sort involved in some instances of concrete exhaustion. It is arguably part of the very meaning of 'dancing' that if one engages in B-type movements (which we are supposing are dancing-intention driven and also not too dissimilar in movement to paradigm cases of dancing), then dancing, even if bad dancing, has thereby occurred. Likewise, it is arguable that stabbing in a death-producing way is by definition an instance of killing. However, there are other brands of what might plausibly be called concrete exhaustion where the necessitation is not purely conceptual. For instance, it might be that Hank's being chased by a mammal on some occasion is concretely exhausted by his being chased by a dog, even though being chased by a dog is arguably only synthetically sufficient for being chased by a mammal. Also, if an instantiation of a mental property were concretely exhausted by the instantiation of some physical property, it seems that the necessitation of the former by the latter would be no more than metaphysical necessitation.[12] So let's interpret 'makes it the case that' as 'is metaphysically sufficient for'.

And regarding the talk of being *required* for the instantiation of some property, let's construe this as referring to *synchronic* dependence, with a dependence of the *sine qua non* variety. So, for example, to say that a set S of concrete items is

[12] Although, in the mental/physical case, the metaphysical necessitation involves the laws of nature (the physical laws in particular). It is metaphysically necessary, given the physical laws, that the same distribution of physical properties is accompanied by the same distribution of mental properties. We will consider how the physical laws play a role in the concrete exhaustion of the mental by the physical later in this section.

required for an instantiation of B is to say that S is a set of concrete items at the time without which B would not have been instantiated.[13]

Of course, for any instantiation of B or B-, there isn't just one set of concrete items on which it depends. There is the set that contains the complete collection of concrete items on which the property instantiation depends. Then there is a set that contains all of those items minus one, and a set containing all of that minus a few more, etc. The set to consider in our explication of *being nothing more than* should be maximal or complete in that it contains *all* of the concrete items on which the instantiation of the property synchronically depends. Otherwise, we might choose as the set of concrete items required for the instantiation of B the set that contains only what is required for the instantiation of B-. Since this smaller set does not suffice for the instantiation of D, we would then arrive at the incorrect result that the instantiation of D is *not* concretely exhausted by the instantiation of B. At the same time, the set should contain no more than what is required for the instantiation of the property. The totality of all concrete items that exist or obtain includes what is required for the instantiation of B-, and this very large set does metaphysically suffice for the instantiation of D. So if we focus on this set as the set required for the instantiation of B-, then we risk the incorrect result that the instantiation of D is concretely exhausted by the instantiation of B-. Thus, while the set of concrete items to consider when deciding what is required for the instantiation of a property should be *complete* (containing all concrete items required for the property instantiation), it should also be *minimally* complete (containing nothing other than those essential items).

To incorporate these qualifications, let's refine the formulation above as follows:

> An instantiation of F at time t is *concretely exhausted* by an instantiation of G at t = (i) there is a set S of concrete items at t that is metaphysically sufficient for the instantiation of F, (ii) there is a minimally complete set, S*, of concrete items at t without which the instantiation of G would not have obtained, and (iii) S is a subset (perhaps a proper subset) of S*.

The notion of concrete exhaustion borrows from Shoemaker's idea that there is something about the realized properties that is a *subset* of something about the realizers—not a subset of *causal powers*, but a subset of *concrete items* involved. However, unlike Shoemaker's subset account, the idea of concrete exhaustion does not require that properties are individuated by their causal powers. As mentioned in Sect. 3.4, while Shoemaker does endorse a strong causal account of properties, according to which, "for each property there is a causal profile that it has in every possible world in which it can be instantiated, and which is such that having that causal profile is sufficient for being that property" (2007, p. 5), his account of realization presupposes only the weaker thesis that each property of a concrete thing "is individuated by a causal profile in the sense that it and it alone has that profile in the actual world and worlds nomologically like it" (p. 142). The presence of

[13] And I leave it completely open here which theory of counterfactuals is most suitable to capturing the truth-conditions of the *sine qua non* claim.

concrete exhaustion as defined above entails neither the weak nor the strong view of property individuation. And this is as it should be if the notion of concrete exhaustion is used to define physicalism, for physicalism itself does not require even a weak account of property individuation.

Suppose, then, that we define physicalism in terms of concrete exhaustion. Suppose we define physicalism as the view that

(CPE) For any mental property M, and any instantiation of M at a time t, there is a physical property or collection of physical properties, P, and an instantiation of P at t such that the former instantiation is concretely exhausted by the latter,

where 'CPE' abbreviates '*Principle of Concrete Physical Exhaustion*'.[14]

It does seem that all physicalists would (or should) accept CPE. Assuming that non-reductive physicalism is a coherent position, one can qualify as a physicalist while rejecting psychophysical type-identity claims. By not requiring that mental properties are physical properties, CPE is open to non-reductive physicalists. At the same time, CPE allows that mental properties are physical, which is just as it should be since physicalists certainly need not deny type-identity.[15] CPE is also compatible with the truth of a token-identity thesis. Recall that Pereboom's account of material constitution (discussed in Sect. 3.5) entails that the tokens constituted are not identical with the tokens that do the constituting.[16] Yet, clearly, one can endorse token-identity claims without thereby forfeiting the label 'physicalist'. At the same time, with no mention of identity in CPE, the definition does allow a physicalist to agree with Pereboom that the relation between mental tokens and physical tokens is one of constitution and not identity. (The definition of what it is to concretely exhaust does not require that the F-instantiation is identical with the G-instantiation, and it does not require that the object or objects involved in the F-instantiation are identical with the objects that participate in the G-instantiation.) Moreover, CPE says nothing about whether mental properties are *functional* properties, which does seem to be a merit since one can qualify as a physicalist while believing that some, perhaps many, mental properties are not functional in any significant sense (believing, e.g., that some mental properties are intrinsic features).

It appears, then, that CPE does not entail anything a physicalist is forced to reject. But is CPE a thesis that only a physicalist would (or should) accept? Suppose that some brand of substance dualism were correct. Then the instantiation of M would involve concrete items (e.g., souls and their activities or conditions) that are wholly

[14] Obviously, the label 'Principle of Concrete Physical Exhaustion' is inspired by the *Principle of Physical Exhaustion* that Hellman and Thompson (1975) offer, which was described in Sect. 2.2. Although, as explained here, CPE is more physically exhaustive than their Principle of Physical Exhaustion.

[15] Indeed, it's arguable that the type-identity thesis is physicalism at its most robust.

[16] This is entailed thanks to his condition (c), which reads: "possibly, y exists at t and it is not the case that y is made up of and materially coincident with x in D at t" (2011, p. 140).

immaterial—wholly immaterial in the sense of lying outside the physical universe. In that case, the collection of concrete items sufficient for M's instantiation includes more than the (minimally complete) set of concrete items required for the instantiation of any physical property. So substance dualism is clearly incompatible with CPE.

However, substance dualists aren't the only ones who reject physicalism. A property dualist can also reject physicalism. Would they join substance dualists in rejecting CPE? To see that they would (or should), note that CPE entails more of what physicalism requires than either the standard token-identity claim, TKI, or the Principle of Physical Decomposition (PD) mentioned in Sect. 2.2.[17] Recall that according to TKI, every instance of a mental property is an instance of a physical property, and according to PD, each mental particular (each instance of and each bearer of a mental property) is either a physical item or possesses a decomposition into parts, all of which are physical items. While both TKI and PD entail that instances of mental properties are instances of physical properties, neither entails that mental properties are instantiated solely in virtue of the instantiation of physical properties. Both allow that while a mental episode is ultimately fully comprised of the stuff of basic physics, what makes it the type of mental episode it is (e.g., what gives it its qualitative or intentional character) depends on more than what physically obtains. Thus, neither TKI nor PD guarantees that mentality is physically exhausted to quite the degree that physicalism demands.

Now consider the degree of physical exhaustion that CPE requires. Suppose that the qualitative character of some mental event depends on more than the instantiation of any physical features. In that case, there is something more than physical features involved. Of course, an additional property is involved (assuming that mental properties are not physical). But that is not all that's additional. It is reasonable to suppose, in general, that which contingent properties (including mental properties) are instantiated is a function of the concrete phenomena that obtain. It seems there would be a difference in mental properties instantiated only with some difference in concrete phenomena, e.g., a difference between the instantiation of pain and the instantiation of pleasure only with some difference in the concrete objects, processes, or conditions involved.[18] If this is correct, and if the qualitative character of some mental event is due to more than the physical features instantiated, then the set of concrete items that suffice for the instantiation of that qualitative feature includes more than what's required for the instantiation of any of those physical properties. It follows that the instantiation of that qualitative feature

[17] PD is the "Principle of Physical Exhaustion" of Hellman and Thompson (1975) when applied to mentality. 'Physical Decomposition' is a more fitting description since as explained here the principle does not guarantee that mentality is physically *exhausted*.

[18] Note, however, that the principle, "no difference in properties instantiated without a difference in concrete phenomena," is not true when the properties are *necessary* features and the items exemplifying them are *abstract*. For instance, there is a difference between the number 3's *being less than 5* and the number 7's *being greater than 5*, but this difference in properties does not require any corresponding difference in concrete phenomena.

is not concretely exhausted by the instantiation of those physical features, which is something CPE does not allow.

So CPE captures more of what physicalism requires than either TKI or PD. But does it capture everything the doctrine demands? As explained in Chap. 2, physicalism seems to entail the truth of a supervenience thesis, according to which, the distribution of physical properties/relations together with the physical laws necessitates the distribution of mental properties/relations. If this physical necessitation did not obtain, then mental phenomena would not be solely a function of physical phenomena, contrary to physicalism.[19]

It is not hard to see that this physical necessitation is secured by CPE. Consider some instantiation of a physical property, P, and the (minimally complete) set S* of concrete items on which it depends, i.e., without which the instantiation of P on that occasion would not obtain. Which collection of concrete items is required on that occasion for P's instantiation would seem to depend on the laws of nature that obtain; with different laws of nature, a different set of concrete items might suffice for the instantiation of P, and in that case, P might be instantiated without S*. Given that P is a physical property, the laws of nature that regulate whether P is or is not instantiated are the physical laws. So, it seems, whether it is true that without S*, the instantiation of P would not have occurred is a matter of the physical laws that obtain. Suppose, then, that world w* contains all of the physical laws that actually obtain. And suppose that P is instantiated in w*. Then since S* is required for P (given the physical laws), S* obtains in w*. Now suppose that the instantiation of M is concretely exhausted by the instantiation of P. Then there is a set, S, of concrete phenomena that guarantees, across all possible worlds, the instantiation of M, and S is a subset of S*. Since S* obtains at w*, and S is a subset of S*, S obtains at w*. It follows that the instantiation of P at w* is accompanied by the instantiation of M.

In general, if CPE is true, then for any instantiation of a mental property, there is an instantiation of a physical property, such that the minimally complete set of concrete items required (given the physical laws) for an instantiation of the latter contains all that suffices for the instantiation of the former. So if CPE is correct, then the actual distribution of physical properties plus the physical laws that obtain guarantees the set of concrete phenomena sufficient for the actual distribution of mental properties.

So it seems that with CPE we have an answer to the threat to non-reductive physicalism that I labeled '***II***'. Let's recall the issue. If the mental details were necessitated by the physical details only with the help of irreducibly psychophysical laws, then mentality would not be solely a function of the way the world is physically. What physicalism requires is that the physical details together with the purely physical laws guarantee all the mental details (as N_p and N_p* mentioned in

[19] As indicated in Sect. 2.1, to allow for the possibility of mental *extras* (possible physical duplicates of the actual world that contain all of the actual mental episodes plus a few additional ones), we might wish to weaken the necessitation claim by insisting only that the *positive* mental facts must remain the same with the duplication of physical facts. In Sect. 2.1 there was also brief mention of some other ways to allow the possibility of mental extras that philosophers have pursued.

Sect. 2.3 entail). Now if mental properties are identical with physical properties, then the necessitation of the mental by the physical is automatically guaranteed, since any property trivially necessitates itself. However, if mental properties are not physical, then it is not at all clear why we should expect this physical necessitation. If mental properties are not physical, then there is the worry, described in Sect. 2.4, that some physical duplicates of the actual world (duplicating the actual world in every physical respect, including all the same physical laws) are worlds where zombies exist or where certain mental episodes are blocked. Such worlds would seem not only to be logically and conceptually possible, but genuine possibilities (genuinely metaphysically possible)—assuming, that is, that mental properties are not physical. Since physicalism precludes physical duplicates of the actual world mentally diverging from the actual world in these ways, to justify their commitment to physicalism, non-reductive physicalists owe a reason for thinking that ***II*** is false —a reason for thinking it is false that

(***II***) mental properties not being identical with physical properties prevents the physical facts from necessitating the mental facts in the way that physicalism requires.[20]

CPE helps us see why ***II*** is false. Perhaps physicalism does not require that mental properties are physical. But it does require a *nothing more than* relation between the mental and the physical, and when that relation is analyzed in terms of concrete exhaustion, as described above, we can understand what the various accounts of realization fail to adequately reveal—namely, how it could be that the physical phenomena necessitate the way the world is mentally even if mental properties are not physical. I have, of course, given no reason to think that CPE is actually true, and therefore no reason to think that physicalism is true. The point is that CPE provides a way of seeing how physicalism could be true (with the necessitation of the mental by the physical that it demands) even if mental properties are not physical.

So CPE provides a way to respond to ***II***. CPE shows how it could be that the mental is necessitated by the physical in the way that physicalism demands even if mental properties are not physical properties. However, with CPE there is another response to ***II*** available that one might find attractive. As described in the next section, the notion of concrete physical exhaustion suggests two plausible and robust definitions of the word 'physical' as applied to properties, definitions according to which mental properties can qualify as physical even if they are not identical with properties of physics or the other natural sciences. According to this response to ***II***, while mental properties are not physical in one sense—the sense of

[20] As mentioned in Sect. 2.4, ***II*** is a different issue from the concern (which I labeled '***I***') that a mere supervenience thesis leaves the nature of the psychophysical determination unexplained. With a supervenience thesis that entails N_P (or N_P*) the covariance of mental properties with physical properties does have an explanation. The explanation is that the physical laws that actually obtain are what make it the case that physical sameness ensures mental sameness. Why this should be the case if mental properties are not physical is what ***II*** has us question.

being properties of physics or the other natural sciences—they are physical in other plausible and robust senses.

4.4 Physical Properties

The term 'physical', when applied to properties, is often restricted to the properties of physics.[21] Although, sometimes the label is used more loosely to include properties of any of the natural sciences, so that chemical and biological properties also count as physical. Despite the continued controversy regarding multiple realizability, I argued in Chap. 1 (1.8–1.10) that considerations of multiple realizability really do show that mental properties are not physical in either of these senses.[22] However, given the notion of concrete exhaustion described in Sect. 4.3, there are other and plausible senses of 'physical' according to which mental properties do qualify as physical despite the multiple realizability.

Consider the property of being a rock. While being a rock is not a property of physics, it is nonetheless not implausible to consider it a physical property. And it seems the reason why it's not implausible to consider it physical is that something is a rock due to *nothing more than* the instantiation of properties of physics. That is, any instantiation of being a rock is *concretely exhausted* by the instantiation of properties of physics. It's arguable that the properties of being a lump of clay, being a piece of cloth, and being a clump of dirt also qualify as physical; even being a stroll in the park, or being a killing by way of stabbing might arguably be considered physical properties. It's arguable that these qualify as physical, not because they are identical with any properties of physics, but because their instantiation is concretely exhausted by the instantiation of properties of physics, i.e., the concrete items at the time sufficient for these properties to be instantiated include no more than those required for properties of physics to be instantiated.

---

[21] It is a matter of significant debate whether we should focus, for the purpose of defining physicalism, on properties of current physics or some future ideal physics. Consider the issue famously called 'Hempel's dilemma': see Hempel (1980). If 'physical' is understood in terms of current physics, then the doctrine of physicalism is likely to be false since current physics is incomplete in many respects. On the other hand, we might define 'physical' in terms of a future, ideal physics, but since we don't know what future physics has in store, the thesis of physicalism would then lack determinate content. For some important work on this dilemma, see Ravenscroft (1997), Montero (1999), Stoljar (2001, 2010), Melnyk (2003), Dowell (2006), Wilson (2006), and Ney (2008a, 2008b). If one appeals to physics in one's definition of 'physical', then for the definition to be complete one will need to decide which physics is being referred to (current, future, ideal,. . .), and one will need to make that decision in a way that eases the threat of Hempel's dilemma. Since I will not undertake that project here, I admit that the definitions of 'physical' I offer here are far from complete.

[22] The arguments based on either the metaphysical or the physical possibility of *immaterial* realizers presented in Sect. 1.8 were shown to be especially effective in establishing non-identity.

There is, then, a rather plausible sense of the word 'physical' applied to properties, according to which, a property is physical just in case its instantiation is concretely exhausted by the instantiation of properties of physics. To avoid confusion, let's introduce the label 'physical*', where

> F is a physical* property = for any instance x of F at a time t, there is a property or collection of properties, P, of physics, and an instance y of P, such that x's instantiation of F at t is concretely exhausted by y's instantiation of P at t.[23]

A physical* property need not itself belong to physics or to any of the other natural sciences. So even if mental properties are not identical with those of the natural sciences, as considerations of multiple realizability seem to show, they can still qualify as physical*.[24]

Still, to hold that mental properties are physical* is to endorse a fair amount. The claim that mental properties are physical* is certainly not one that any substance dualist would accept. Substance dualists would deny that the instantiation of mental properties is concretely exhausted by the instantiation of properties of physics, insisting that immaterial concrete items (e.g., souls and their immaterial conditions) are required. Non-physicalists who reject substance dualism would also deny that mental properties are physical*. Property dualists who reject physicalism believe that while a mental event occurs in physical space, what makes it the type of mental event it is (e.g., what gives it its qualitative or intentional character) is a matter of more than the features of physics it has.[25] As explained in Sect. 4.3, this claim that the mental is something more entails that the instantiation of a mental property is not concretely exhausted by the instantiation of any physical properties, where these "physical properties" clearly include properties of physics. Since the claim that mental properties are physical* is one that all non-physicalists would reject, the notion of being physical* does count as a rather robust sense of 'physical' when applied to properties.

There might be some who go by the label 'physicalist' but would still reject the claim that mental properties are physical*. I have in mind someone who extends the label 'physical' to properties of any of the natural sciences, and who also believes that mentality is solely a function of biochemical or neurological properties, but for some reason denies that the instantiation of any organic property is concretely exhausted by the instantiation of properties of physics. A physicalist of this sort

---

[23] The choice is open whether to rely on actual current physics, or something sufficiently similar to current physics, or to appeal to some future and/or ideal physics.

[24] Moreover, the definition does *not* require that physical* properties are *necessarily* physical*. Instantiations of being a rock are in fact concretely exhausted by the instantiation of properties of physics, which allows that there are possible worlds where rocks are comprised of immaterial substances of some sort. This metaphysical possibility does not make it any less plausible to call the property of being a rock 'physical'.

[25] Use of the expression 'property dualists who reject physicalism' seems to imply that property dualists can accept physicalism, and of course they can if non-reductive physicalism is a coherent position.

might deny that mental properties are identical with properties of biochemistry and neuroscience, but still believe that mental properties are physical**, where

> F is a physical** property = for any instance x of F at a time t, there is a property or collection of properties, P, belonging to the natural sciences, and an instance y of P, such that x's instantiation of F at t is concretely exhausted by y's instantiation of P at t.

This is a weaker sense of 'physical', but it seems strong enough that it, too, would be rejected at least by the typical opponent of physicalism. Substance dualists would certainly deny that the instantiation of mental properties is concretely exhausted by the instantiation of properties of any of the natural sciences. Non-physicalists who reject substance dualism are also likely to deny that mental properties are physical**. Property dualists who reject physicalism generally believe that, at least for some mental processes, what gives the process its special mental character is a function of more than the properties of physics it has, and also a function of more than its chemical or biological features. So the claim that mental properties are physical** seems to be one that non-physicalists in general would reject. Thus, while not as robust as the notion of being physical*, the notion of being physical** is hardly an anemic sense of 'physical'.

As explained in Sect. 4.3, the notion of concrete exhaustion reveals why ***II*** is false, thereby showing how one can remain faithful to physicalism even while denying that mental properties are physical. And yet, the notion of concrete exhaustion succeeds in this way only by revealing two plausible senses of 'physical', and one can believe that mental properties are physical in either sense (*physical** or *physical***) without believing that they are identical with any properties of the natural sciences. It seems, then, that we have a second line of response to threat ***II***. We can reply that if physicalism is correct, then there is a dependency of the mental on the physical (which consists in the relation of concrete exhaustion), and this dependency is strong enough that mental properties can still qualify as physical in a plausible and significant sense even if they are not identical with properties of the natural sciences.

4.5 Conclusions

Let's close by taking stock of what has been established in this chapter. It was shown in Sect. 4.3 that the appeal to concrete exhaustion captures the physicalist intuition that an instantiation of a mental property on some occasion is nothing more than the instantiation of physical properties. So, it seems,

(i) accepting CPE is enough to warrant the title 'physicalist'.

Accepting CPE does not require believing that mental properties are identical with any of those of the natural sciences. This fact together with (i) entails that

(ii) qualifying as a physicalist does not require believing that mental properties are identical with those of the natural sciences.

In particular, CPE reveals why ***II*** is false, since

(iii) CPE shows how it could be that the mental facts are necessitated by the physical facts in the way that physicalism requires even if mental properties are not physical.

There is also the fact, emphasized in Sect. 4.3, that CPE seems neutral enough on various metaphysical issues to be a thesis that any physicalist would (or should) accept. So it also seems that

(iv) accepting CPE is necessary for being a physicalist,

and from (i) and (iv) it follows that

(v) physicalism is plausibly viewed as the claim that CPE is true.

It was argued in Sect. 4.4 that the notion of concrete exhaustion suggests two senses of 'physical' (*physical** and *physical***) when applied to properties, each of which allows that mental properties are physical even if they are not identical with properties of physics or any of the other natural sciences. Moreover, these are robust senses of 'physical'; regarding physical*, it seems only a physicalist would believe mental properties are physical in that sense. Thus,

(vi) there are plausible and robust senses of 'physical', and mental properties can qualify as physical in these senses without being properties of physics, chemistry, or biology.

But now recall Chap. 1 where I argued that one should be a non-reductionist—in the sense of denying that mental properties are identical with properties of the natural sciences. Given (ii), one can qualify as a non-reductionist in this sense while also being a physicalist. Although, physicalists in general would believe that mental properties are either physical* or physical**. So now it seems that

(vii) while some physicalists might be "non-reductive," there is another significant sense in which they are reductionist, i.e., by believing that mental properties are identical with physical* or physical** properties.

References

Baker, L. 2013. Pereboom's robust nonreductive physicalism. *Philosophy and Phenomenological Research* 86: 736–744.

Chalmers, D. 1996. *The conscious mind*. New York: Oxford University Press.

Correia, F. 2008. Ontological dependence. *Philosophy Compass* 3: 1013–1032.

Correia, F. 2010. Grounding and truth-functions. *Logique et Analyse* 53: 251–279.

Davidson, D. 1969. The individuation of events. In *Essays in honor of Carl G. Hempel*, ed. N. Rescher, 216–234. Dordrecht: Reidel.

Dowell, J. 2006. The physical: Empirical, not metaphysical. *Philosophical Studies* 131: 25–60.

Fine, K. 1994. Essence and modality. *Philosophical Perspectives* 8: 1–16.

Fine, K. 1995. Ontological dependence. *Proceedings of the Aristotelian Society. New Series* 95: 269–290.

Fine, K. 2012. Guide to ground. In *Metaphysical grounding: Understanding the structure of reality*, ed. F. Correia, and B. Schnieder, 37–80. Cambridge: Cambridge University Press.

Hellman, G., and F.W. Thompson. 1975. Physicalism: Ontology, determination, and reduction. *The Journal of Philosophy* 72: 551–564.

Hempel, C. 1980. Comments on Goodman's *Ways of worldmaking*. *Synthese* 45: 193–199.

Jackson, F. 1982. Epiphenomenal qualia. *Philosophical Quarterly* 32: 127–136.

Jackson, F. 1986. What Mary didn't know. *The Journal of Philosophy* 83: 291–295.

Jackson, F. 2003. Mind and illusion. In *Minds and persons*, ed. A. O'Hear, 251–271. Cambridge: Cambridge University Press.

Jackson, F. 2012. In defence of the identity theory mark I. In *New perspectives on type identity: The mental and the physical*, ed. S. Gozzano, and C. Hill, 150–166. Cambridge: Cambridge University Press.

Kim, J. 1966. On the psycho-physical identity theory. *American Philosophical Quarterly* 3: 227–235.

Quine, W.V.O. 1950. Identity, ostension, and hyposthasis. *Journal of Philosophy* 47: 621–633.

Quine, W.V.O. 1985. Events and reification. In *Actions and events: Perspectives on the philosophy of Donald Davidson*, ed. E. Lepore, and B. McLaughlin, 162–171. Oxford: Basil Blackwell.

Melnyk, A. 2003. *A physicalist manifesto*. Cambridge: Cambridge University Press.

Montero, B. 1999. The body problem. *Nous* 33: 183–200.

Ney, A. 2008a. Physicalism as an attitude. *Philosophical Studies* 138: 1–15.

Ney, A. 2008b. Defining physicalism. *Philosophy Compass* 3: 1033–1048.

Pereboom, D. 2011. *Consciousness and the prospects of physicalism*. New York: Oxford University Press.

Ravenscroft, I. 1997. Physical properties. *The Southern Journal of Philosophy* 35: 419–431.

Rosen, G. 2010. Metaphysical dependence: Grounding and reduction. In *Modality: Metaphysics, logic, and epistemology*, ed. B. Hale, and A. Hoffmann, 109–135. Oxford: Oxford University Press.

Schaffer, J. 2012. Grounding, transitivity, and contrastivity. In *Metaphysical grounding: Understanding the structure of reality*, ed. F. Correia, and B. Schnieder, 122–138. Cambridge: Cambridge University Press.

Shoemaker, S. 2007. *Physical realization*. New York: Oxford University Press.

Stoljar, D. 2001. Two conceptions of the physical. *Philosophy and Phenomenological Research* 62: 253–281.

Stoljar, D. 2010. *Physicalism*. London: Routledge.

Wilson, J. 2006. On characterizing the physical. *Philosophical Studies* 131: 61–99.

FSC
www.fsc.org
MIX
Papier aus verantwortungsvollen Quellen
Paper from responsible sources
FSC® C105338

If you have any concerns about our products,
you can contact us on
ProductSafety@springernature.com

In case Publisher is established outside the EU,
the EU authorized representative is:
Springer Nature Customer Service Center GmbH
Europaplatz 3, 69115 Heidelberg, Germany

Printed by Libri Plureos GmbH
in Hamburg, Germany